Lecture Notes in Computer Science

T0237813

Commenced Publication in 1973
Founding and Former Series Editors:
Gerhard Goos, Juris Hartmanis, and Jan van Leeuwen

Editorial Board

Jörn Altmann   Daniel J. Veit (Eds.)

Grid Economics and Business Models

4th International Workshop, GECON 2007
Rennes, France, August 28, 2007
Proceedings

 Springer

Volume Editors

Jörn Altmann
Seoul National University
Seoul 151-744, South Korea
E-mail: jorn.altmann@acm.org

Daniel J. Veit
University of Mannheim
Mannheim, Germany
E-mail: veit@iw.uka.de

Library of Congress Control Number: 2007932974

CR Subject Classification (1998): C.2, K.4.4, H.3, J.1

LNCS Sublibrary: SL 5 – Computer Communication Networks and
Telecommunications

ISSN   0302-9743
ISBN-10   3-540-74428-2 Springer Berlin Heidelberg New York
ISBN-13   978-3-540-74428-3 Springer Berlin Heidelberg New York

Springer is a part of Springer Science+Business Media

springer.com

© Springer-Verlag Berlin Heidelberg 2007
Printed in Germany

Typesetting: Camera-ready by author, data conversion by Scientific Publishing Services, Chennai, India
Printed on acid-free paper   SPIN: 12112494   06/3180   5 4 3 2 1 0

GECON - Grid Economics and Business Models

Although more and more applications can be executed on the Grid and the management of resources shifts from "buying hardware" to "computing on demand", a global commercial Grid system does still not exist. The concurrent execution of Grid applications on this commercial Grid system would allow the emergence of new marketplaces, namely, markets for processing power, storage, bandwidth, and information services. The emergence of such markets would be the first step towards achieving the goal of a global, commercially operated, and efficiently utilized Grid system. These markets, which can be used for resource selection, resource allocation, access control, and resource planning, are addressed in the research domain of Grid economics. Their realization will make the Grid evolve from a large, distributed system offering best-effort, zero-cost computing services into an environment capable of resolving differences in service preferences of Grid users. This new economic-enhanced Grid will create additional value for its participants through better means for expressing preferences, sharing resources, and generating revenue. It builds a stable and scalable economy of resources.

In such a new Grid system, there is no longer a distinction between specific Grid users. The general public, academic institutions, SMEs or larger institutions are able to use the services of such a new commercial Grid system in the same way. The concrete services could range from a simple service, charged at low rate but aiming for mass usage, to complex services encapsulating e-Science or HPC methods. However, in order to make this vision operate in practice, innovative schemes of charging for the usage of services have to exist in this novel kind of service-oriented, market-based Grid system, stimulating providers to offer and customers to demand these services.

The 4th International Workshop on Grid Economics and Business Models, GECON 2007, aimed at presenting current results and innovative research in the area of Grid economics. In particular, the target is to facilitate the discussion of new business models for the Grid and the capability of the existing Grid to allow the economic operation of the Grid. The purpose of this endeavor is to concretize directions of research and amendments to existing technologies, aiming at the successful deployment of a global commercial Grid system.

In the first contribution, Thanos et al. identify the main objectives and associated economic issues while applying Grid in business purposes. Herein, the authors state that real-life economics within business perspectives is more important than complex theoretical economic problems. The concept picks up at players, who do not exhibit the required technological expertise in order to elaborate on Grid issues. Thanos et al. aim at identifying critical economic issues that should be taken into consideration by industrial partners in order to create trust and confidence in this novel technology.

Hwang and Park investigate the adoption of enterprises to autonomous Grid computing rather then individually owned ICT infrastructures. They highlight the importance of the quality of Grid solution providers for this decision based on a conjoint analysis. Their results show that interoperability is the most indispensable element to a successful utilization of Grid infrastructures in enterprises.

In the third contribution, Altmann and colleagues formulate a taxonomical approach to Grid business models. They survey the development and origin of Grid technologies and focus on the importance of business-directed values when trying to commercialize today's Grids. Therein, they identify the reduction of costs, the improvement of efficiency, the creation of novel products and services as well as the quality and collaboration between companies as key factors for the differentiation of Grid business models. The paper concludes by applying the proposed taxonomy to a utility computing scenario and a software-as-a-service scenario in practice.

Stanoevska-Slabeva and Zsigri propose a generic value chain for the Grid industry. In their contribution, they suggest a case study on aggregating results from different Grid middleware modules into a generic Grid value chain.

In their contribution, McKee and coauthors propose a set of strategies for acting in future service-oriented markets. The costs of negotiations are put in relation to the value of the offer under negotiation. Hence, the contribution adds to the state of the art by extending the vision of service level agreements (SLAs) within service frameworks.

Sandholm and Lai propose a novel, prediction-based enforcement of performance contracts. Their approach aims at controllable quality of service (QoS) within Grid computing platforms. The proposed mechanism is based on a hybrid resource allocation system using both proportional shares and reservations.

In the seventh contribution, Huang proposes a flexible, refundable auction concept for limited capacity suppliers. The mechanism that is introduced is called Decreasing Cancellation Fee Auction (DCFA). It proposes the use of uncertainties of the resource availability for the substantiation of the consumers' decision to use resources. A partial refund of the users payment for reservations provides an incentive to participate in the market. It shows that the mechanism is incentive compatible, individually rational at still high efficiency.

The following contribution by Vanmechelen and Broeckhove aims at introducing a dynamic pricing scheme by comparing single-unit Vickrey auctions and commodity markets. They highlight that a key research issue is the choice of a market organization. The results that the authors provide are a quantitative analysis of the comparison between the two indicated allocation schemes. Based on their simulation results, they conclude that – although similar outcomes are achieved – a commodity market organization leads to more stable market behavior at the cost of higher communicative requirements.

Giordano and Di Napoli focus on provisioning a sophisticated computing methodology in order to provide Grid services in a continuous and seamless way. The main contribution is the provisioning of a flexible and easily programmable middleware to experiment with different economy-driven scheduling policies for service-oriented computing.

In the tenth contribution, Franke and coauthors address necessary modifications and extensions to existing Grid computing approaches in order to meet modern business demand. They attempt to bridge the gap between architectures for solving large scientific problems and concepts for dealing with performance, monitoring aspects, security, and isolation issues.

Assunção et al. elaborate on the simulation of service-oriented computing and provisioning policies for autonomic utility Grids. They address the issue of QoS in the context of the provider's decision about the resource allocation. In their simulations, they propose a decentralized, self-organizing resource allocation and provisioning scheme based on Friedrich A. von Heyek's Catallaxy approach.

In the next paper, Maillé and Toka present a peer-to-peer backup system where users offer some of their storage space to provide services for others. Their economic model differentiates from the regular peer-to-peer models by incentivizing users to contribute to a service. In the following, they show that their proposed symmetric scheme is outperformed by a revenue maximizing monopoly with respect to social welfare maximization.

Finally, contributions on the research projects ArguGrid, AssessGrid, CatNets, edutain@grid, GridEcon, and SORMA give an overview on current and ongoing research in Grid economics.

In preparation of this fourth workshop, 96 reviews were written for which we would like to thank our Program Committee. The Program Committee served within a very short time frame in order to enable the successful preparation of this workshop. In particular, we would like to thank for this: Hermant K. Bhargava, Rajkumar Buyya, John Chuang, Costas Courcoubetis, Dang Minh Quan, John Darlington, Torsten Eymann, Thomas Fahringer, Kartik Hosanager, Chun-Hsi Huang, Junseok Hwang, Harald Kornmayer, Ramayya Krishnan, Kevin Lai, Hing-Yan Lee, Jysoo Lee, Steven Miller, Dirk Neumann, David Parkes, Omer Rana, Peter Reichl, Simon See, Satoshi Sekiguchi, Burkhard Stiller, Yoshio Tanaka, Maria Tsakali, Bruno Tuffin, Gabriele von Voigt, Kerstin Voss, and Stefan Wesner. As a result of the review process, the overall acceptance rate of the workshop was at 40% of the submitted contributions.

Furthermore, we would like to thank Alfed Hofmann and Ursula Barth from Springer, who made this volume possible through their patience and continuously positive support. We would also like to thank the organizers of the 2007 Euro-Par Conference – namely, Luc Bougé – for the substantial support in hosting the GECON 2007 workshop in Rennes, France.

Finally, we would like to express our gratitude to Sonja Klingert for her extensive effort in preparing the manuscripts for the proceedings of this workshop.

August 2007

Daniel J. Veit
Jörn Altmann

Organization

GECON 2007 was organized by the International University in Germany, Bruchsal, Germany, Department of Computer Networks and Distributed Systems and the University of Mannheim, Germany, Department of Business Administration and Information Systems, in collaboration with EuroPar 2007.

Executive Committee

Chair   Jörn Altmann (Intl. University in Germany, Germany)
Vice Chair   Daniel Veit (University of Mannheim, Germany)
Organization Chair   Sonja Klingert (Intl. University in Germany, Germany)

Program Committee

Hermant K. Bhargava (UC Davis, USA)
Rajkumar Buyya (University of Melbourne, Australia)
John Chuang (UC Berkeley, USA)
Costas Courcoubetis (Athens University of Economics and Business, Greece)
Dang Minh Quan (International University, Germany)
John Darlington (Imperial College, UK)
Thomas Fahringer (University of Innsbruck, Austria)
Kartik Hosenager (University of Pennsylvania, USA)
Chun-Hsi Huang (University of Connecticut, USA)
Junseok Hwang (Seoul National University, Korea)
Ramayya Krishnan (Carnegie Mellon University, USA)
Kevin Lai (HP Labs, USA)
Hing-Yan Lee (National Grid Office, Singapore)
Jysoo Lee (KISTI, Korea)
Steven Miller (Singapore Management University, Singapore)
Dirk Neumann (Karlsruhe University, Germany)
David Parkes (Harvard University, USA)
Omer Rana (Cardiff University, UK)
Peter Reichl (Telecommunications Research Center Vienna, Austria)
Simon See (Sun Microsystems, Singapore)
Satoshi Sekiguchi (AIST, Japan)
Burkhard Stiller (University of Zurich, Switzerland)
Yoshio Tanaka (AIST, Japan)
Maria Tsakali (European Commission, Belgium)
Bruno Tuffin (IRISA/INRIA, France)
Gabriele von Voigt (University of Hannover, Germany)
Kerstin Voss (University of Paderborn, Germany)
Stefan Wesner (HLRS, Germany)

Sponsoring Institutions

International University in Germany, Bruchsal, Germany
University of Mannheim, Mannheim, Germany
Springer LNCS, Heidelberg, Germany
EuroPar 2007, Rennes, France

Table of Contents

Research Projects on Grid Economics

Adopting the Grid for Business Purposes: The Main Objectives and the Associated Economic Issues

George A. Thanos, Costas Courcoubetis, and George D. Stamoulis

Network Economics and Services Laboratory, Department of Informatics
Athens University of Economics and Business, Patision 76, Athens 104 34, Greece
{gthanos,courcou,gstamoul}@aueb.gr

Abstract. Grid technology offers numerous opportunities for the players involved. Despite the fact that the academic community has already exploited many of them, there is an evident reluctance from the business community to act likewise. Recent analysis reveals that the problem lies in overcoming certain business barriers rather than technological ones. At this stage understanding the real-life economic issues from a business perspective is deemed as more important than gaining understanding of complex theoretical economical problems, such as those related to accounting or resource sharing mechanisms especially in cases where the players do not exhibit the required technological expertise. This paper is stimulated from interaction with players from the industry and aims to fill this gap. In particular, we identify and evaluate a number of economic issues that should be taken into consideration by industrial players so that their trust and confidence in the adoption of this promising technology be increased.[1]

Keywords: Grid Technology, economics, resource sharing, virtual organisations, market, economies of scale, network externalities etc.

1 Introduction

Grid technology promises a new way of delivering services across IP-based infrastructures. These range from common ones, such as existing mass multimedia services, to more complex and demanding customised industrial applications. Over the last years Grid technology has proven its merits through enabling the execution of highly resource demanding applications for the scientific community some of which were previously only realised over expensive high-performance computing (HPC) centres.

However, in order for Grid technology to fulfil the aforementioned promise, it has first to be adopted by the diverse business community thus being provided and consequently validated, by a significantly larger number of providers and users. Recent studies [1] and European initiatives [2] have indicated a reluctance and slow take-off of Grid technology and market by the industry, something attributed mainly to economic and market barriers rather than to technological ones.

[1] This project has been partly supported by FP6 EU-funded IST projects BEinGRID (IST5-034702) and GridEcon (IST5-033634).

D.J. Veit and J. Altmann (Eds.): GECON 2007, LNCS 4685, pp. 1–15, 2007.

So far there has being a lot of work around theoretical economical analysis examining issues like accounting and resource sharing mechanisms for Grid architectures etc. However, our experience from interacting with industry players and discussing their concerns has shown that prior to solving complex architectural issues there is an evident need for analysing the Grid phenomenon and its economic side effects from a business perspective. Thus, in this paper we identify and analyse those criteria and economic issues that a new player should take into consideration prior to making his decision whether to adopt Grid technology for his business or not and how these will affect his Grid business afterwards. Such a decision can be made by means of a relevant model, which will take the factors identified in the present paper as inputs. Our overall aim is to increase the confidence of the industry towards Grid adoption by exposing the business issues, both positive and negative ones, that once taken into careful consideration by the value chain players will enable them to realise the numerous opportunities that Grid technology has to offer and at the same time construct feasible business plans to fully exploit them.

Our identification and analysis has been performed with support by the Integrated Project Business Experiments in Grid – BEinGRID [3], European Union's largest integrated project funded by the Information Society Technologies (IST) research, part of EU's Framework Programme 6 [4]. The communication with 18 real-life Business Experiments from various industries provided the practical framework to validate our theoretical analysis.

The paper is structured as follows: a brief introduction to Grid economics is presented in Section 2 followed by a discussion on the main economic objectives for adopting the Grid and an initial identification of associated economic issues in Section 3. Section 4 identifies and analyses a number of economic issues related to Grid adoption from the industry whereas Section 5 provides a case study and evaluation of how these issues affect real-life scenarios. Section 6 provides some concluding remarks.

2 A Brief Introduction to Grid Economics and Related Work

Firstly, it is imperative to review some basic definitions related to Grid Technology and the current work in Grid Economics. To start with, we define a Grid service as a Web Service that provides some well-defined interfaces and follows specific conventions [5]. The interfaces address issues such as address discovery, dynamic service creation, lifetime management, notification, and manageability. The conventions regulate naming and upgradeability of services. Each service described in the Open Grid Services Architecture (OGSA) [6] is a single Grid service or a composition of Grid services. A Grid middleware is typically composed of several Grid services with different functionality. Usually, at least the following functionalities are covered: resource management, Job management, Service discovery, scheduling, accounting and security.

Nowadays, a single business process and value chain of a company can be performed by several business partners. The company involved in this process is then a virtual company or organization (VO), as it is only a temporary aggregation of partners in order to perform a specific process. The corresponding concept from economics is called the coalition. VOs can be seen one of the most important drives for Grid technology adoption as it allows these organisations to efficiently share and utilise their geographically distributed computing, storage and data resources over a common infrastructure.

Among the first to raise a number of true economic issues focused on the commercialization of Grid resources (i.e. computing) were Kenyon and Cheliotis. Specifically, in their work they argue that Grid commodity is rather a stochastic one rather than as a deterministic one (such as oil, electricity, etc). Since Grid resources are non-storable, the authors claim that future contracts will be the basic building blocks in Grid environments instead of spot markets. Market uncertainty and decision support are the most important issues that need to be addressed in this context.

The authors identify a set of requirements for commercialization of Grid resources such as product construction and reservation, contract management, clearing, accounting and billing, trading support, price formation and decision support. Also, in [7], Cheliotis et al. set a number of important questions on the successful creation of a Grid market. They argue that the most important part for a successful Grid market creation is to fully understand and foster user requirements and demands. Overall, [8], [9], [7] mostly define the most important issues for Grid commercialization and they do not propose any specific solutions for them.

Gray on the other hand in [10] discusses the economic tradeoffs of doing Grid-scale distributed computing (WAN rather than LAN clusters). Specifically, Gray analyzes the economics of outsourcing. Using simple commercial examples, he calculates the corresponding value of 1$ for bandwidth over the WAN, for number of CPU instructions, for CPU time, for disk space, for database accesses and for disk bandwidth. Identifying communication cost as a bottleneck, Gray concludes on a rule of thumb regarding outsourcing, according to which computations must be nearly stateless and have more than 10 hours of CPU time per GB of network traffic before outsourcing the computation makes economic sense. Otherwise, LAN cluster provide a more economically viable alternative.

Probably the most extensive work on Grid Economics up-to-date has been performed by the GRIDS (Grid Computing and Distributed Systems) laboratory, headed by Buyya. Their most significant research work related to our work is the Economy Grid project where it is clearly identified that a major challenge for next-generation Grid computing is the creation of an "Economy Grid", meaning a competitive realistic Grid Marketplace that regulates supply and demand, and offers the right incentives to players (suppliers and consumers) for improving the utilization of resources. The next step was the Gridbus [11] project, aiming at producing a set of Grid middleware technologies to support e-science and e-business applications. In some of the designed and developed components for this technology one will find incorporated features relevant to "Grid Economics", such as a broker agent software for job scheduling, a market directory for publishing and searching for available services, and a centralized infrastructure that provides accounting and payment services. The "Economy Grid" project, the GRACE architecture and an overview of related work on price setting, market-based resource allocation and scheduling systems are presented in [12].

Other works in the Grid Economics include a centralized strategy-proof architecture for Grid Computing by Egg [13] and the Mojo Economy [14], the Weng

Price-setting mechanisms [15], the price prediction mechanisms by [16], and work driven from European funded IST projects.

As already mentioned the aforementioned work is more focused in the theoretical analysis of economic mechanisms and fails to analyse specific economic issues from a business perspective such as the economies of scale/scope, network externalities, free-riding problems, information asymmetry, and impacts to other markets etc which we will address in the subsequent sections.

3   Economic Objectives for Adopting the Grid and Initial Identification of the Associated Economic Issues

The aim of this section is to discuss the main economic objectives for adopting the Grid for Business and identify the underlying economic issues/concerns. We propose at this stage the main three alternative economic reasons for Grid to be used in commercial applications. By keeping the number of alternatives small and hence abstracting to a significant level the implementation context, we can understand the economics better. These are discussed in the subsequent sections.

3.1   Optimization of Processing Power in a Single Organization

A single organization may require processing power that cannot be provided by means of stand-alone machines. By interconnecting these machines in a Grid, high processing power can be used even by a single application. Thus, the organization achieves both a high peak processing capacity and a high average utilization of the processing power available, since this can be flexibly allocated to multiple Grid-enabled applications. These features also lead to increased cost-efficiency for the infrastructure deployed. This is particularly important for a large organization with several departments scattered around the world, each possessing its own local infrastructure. Interconnecting these in a Grid attains the aforementioned performance enhancement, high exploitation of resources, and cost-efficiency and economies of scale, due to the fact that interconnection of all machines improves utilization of each individual one. Moreover, the whole approach is scalable, due to the fact that the Grid middleware provides automatic load balancing and transparent usage of the hardware. Besides these, if the various departments possess complementary infrastructure, then the organization also attains economies of scope.

Regarding management, since the Grid belongs to single organization, a centralized approach is always an option. On the other hand, particularly if there are multiple departments in the organization, with some notion of autonomy (e.g. own infrastructure contributed to Grid and IT budget), then self-management of the Grid by means of economic/market mechanisms is possible and probably preferable. Indeed, the centralized approach requires complete information, which is not always straightforward to gather in a highly distributed single-domain system. On the other hand, a market mechanism defining prices for accessing and using the Grid resources by the various departments provides the right incentives for rational usage and results in shaping of demand according to the actual needs, which in fact may be thus discovered; prices may either be really monetary, or virtual ones with each department being allocated a Grid virtual budget. This approach also requires accounting functionality, e.g. for monitoring the usage of resources by the various departments and assigning the relevant charges, as well as specification of the right SLAs and appropriate tariffs for them.

3.2 Sharing of Complementary Resources in Multi-provider Environments

Consider a group of organizations, each of which possesses its own resources, which are complementary to each other. For example, organization A possesses a powerful database server, while B has a huge amount of data and C possesses an application running over its server that requires data such as that of B. Clearly, when collaborating in the form of Grid, all organizations can bring together a powerful outcome, while each of them exploits very highly its own resources at a cost-efficient way, without needing to invest to the missing resources that are now contributed by others. In this case, the collaborating organizations enjoy economies of scope, since bringing all resources together by means of Grid broadens their scope of applicability. Apart from serving their own needs by forming a Grid, organizations with complementary resources may also form a Virtual Organization serving third parties. The formation of VOs has a considerable impact to the market; see item 3 below. If the group forming the Grid is not closed, then network externalities and economies of scale may arise in the case where new organizations can join the group, thus enhancing the associated gains per participant.

Regarding self-management, the collaboration of the participants in the Grid should be regulated by means of market mechanisms that provide them with the right incentives to both contribute to the Grid the resources promised and not to free-ride those of the others. For example, a global agreement can prescribe that all contribute as necessary. Similarly to peer-to-peer systems such agreements can be based either on rules prescribing a fixed minimum contribution for all participants or on rules regulating the consumption levels of each participant (quantitatively or qualitatively) in accordance/relation with his contribution over time. These rules should be complemented by accounting functionality that certifies conformance with them. Also, an internal market mechanism, based on SLA and monetary prices for these SLAs can also be employed as an effective approach for self-management, particularly in cases where the level of contribution of the various participants is not symmetric, and a global agreement is hard to be reached. These ideas apply to the case where the Grid is formed in order to serve the participants' own needs, including the case of a single organization with multiple departments (see item 1). If the participants also serve third parties, then the relations between the former and the latter should also be managed by means of market mechanisms.

3.3 Offering Utility Computing Services

This amounts to offering applications (software) and computing services (hardware) on a pay-per-use basis rather than by means of licensing or long term static agreements (leasing, etc.). In this model, applications are sold as components according to the SOA architectural concepts; customers can design their full solution by combining components from different providers and run these on their own premises or again using some Application Service Provider (ASP) computing services. Essentially, this application level Grid allows for a new version of the application based on components to be accessed by the customers. This version is more affordable to infrequent users of the application, who now have a benefit compared to investing on the corresponding software license and/or computational infrastructure. Therefore, both these users and the service provider gain, since this version increases the demand for the service by making it affordable at lower costs. At the lower layers an ASP may benefit from Grid computing services using his own infrastructure complemented with utility computing services offered form third parties. The issues discussed in the previous items regarding high performance, economies of scale and scope etc. are still applicable here.

Nevertheless, other interesting economic issues arise too in the present case. In particular, we now have a new market (that of the pay-per use application), in which: a) the proper SLAs should be offered to customers, and b) resources should be self-managed and the revenue should be properly distributed to the players involved, while c) this market also has significant impact on other markets!

In case where this provider is a single organization, the self-management of its resources is attained through its incentives for optimizing its profits obtained from the market; for example, the predictions for market demand and the revenues foreseen accordingly can serve as an input of a capacity expansion policy. In case where the Grid provider is a virtual organization (or a single one yet with multiple participating departments), then additional self-management mechanisms are needed in order to pass the revenues to the various participants according to their level of contribution.

As already mentioned, the new market created in the present case may have a significant impact to other markets too. In particular, a Small and Medium Enterprise (SME) that cannot afford investing on a license or on infrastructure obtains new capabilities by outsourcing its missing application to the Grid provider on a pay-per-view basis. Thus, such an SME can now serve as a provider in another market, in which this application is a necessary capability for each provider. Therefore, the Grid version of the application leads to a reduction of the barriers of entry in the other market, which is now more competitive. This in turn may have a positive effect to the Grid provider itself, since the SMEs penetration in this new market generates additional demand for the Grid application. If beneficial for the Grid to expand, which is particularly the case if economies of scale and scope apply, then the customer SMEs will benefit even more by the expansion of Grid. Network externalities also apply here.

A summary of how the aforementioned issues impact the Grid adoption decision process is presented in the next table:

Table 1. The impact of the economic issues in the Grid adoption decision process *(1: Strong Influence, 2: Medium Influence, 3: Weak influence)*

Categories/ Adoption Decision Influence	Econ. of Scale	Econ. of Scope	Network External ities	Self-manag ement	New markets	Impact to other markets	Free-riding	Info. Asym metry	Perform. Different iation
Optimisation of processing power	1	2	3	1	2	2	3	3	1
Sharing of complementary resources	1	1	1	1	2	2	2	2	2
Utility Computing	1	2	2	1	1	1	2	3	2

4   Analysing the Economic Issues Associated with Grid Business Scenarios

In the previous section we have briefly identified a number of economic issues that should be taken into account for Grid technology adoption in the business context. Consequently, these identified issues are listed below with a brief explanation of their meaning, their relevance in the context of Grid business scenarios and their significance. Following this analysis in subsequent sections we aim to evaluate them and further discuss their impact in terms of real-life Grid Business scenarios in Section 5. As will be seen there, these issues together with the objectives determine the decision of whether to adopt Grid or not; see Figure 1.

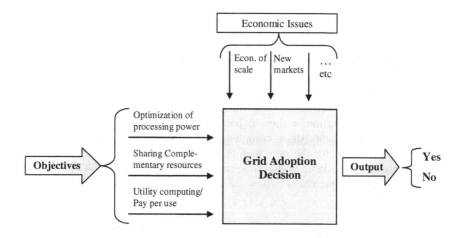

Fig. 1. Grid adoption decision process

Economies of scale and scope (complementarities)
As a definition it can be said that there are economies of scale in production if the cost per unit of production declines with the number of units produced. (Thus, "economies of scale" is a descriptive, quantitative term). Due to economies of scale, larger companies have greater access to markets in terms of selecting media to access those markets, and can operate with larger geographic reach whereas for traditional companies, size does have its limits, where additional size actually increases costs to companies (impacts communications costs etc., diminishing returns).

Economies of scope are conceptually similar to economies of scale. Whereas economies of scale apply to efficiencies associated with increasing the scale of production, economies of scope refer to efficiencies associated with broadening the scope of the service(s) offered, of marketing and distribution thereof etc. That is, while economies of scale refer primarily to supply-side changes (such as level of production), economies of scope also refer to demand-side changes (such as marketing and distribution).

The economic concepts of "economies of scope" and "economies of scale" similarly apply to the Grid market, where the integration of Grid technologies from a value chain actor and the consequent infrastructure and application improvements (e.g. in terms of performance) can lead to a production scale of the company's end-user products. Furthermore, sharing complementary resources among organizations could lead to the specification and the market entrance of new market products, thus to a realization of the "economies of scope" theory.

Network externalities

Network externalities are the effects on a user of a product or service of others using the same or compatible products or services. Positive network externalities exist if the benefits are an increasing function of the number of other users. Negative network externalities exist if the benefits are a decreasing function of the number of other users. For example a positive network externality arises in telephony, when the network expands; thus, each new user has more opportunities to communicate with others, and thus may be the amount that he is willing to pay for subscribing to the network depends on who or how many other parties are connected to it. Such an externality also applies to Internet, together with a negative externality that the more users the higher the congestion.

In a similar fashion, network externalities strongly apply to the Grid case, where for example an organization wishes to participate in a Virtual Organisation (VO) structure whose participants share complementary resources and the final outcome and thus "Grid" benefit for the new member strongly depends proportionally to the amount of resources available at that time to be shared by the other VO participants, i.e., the number of total members.

Self-management issues

Grid environments usually depict strong-collaboration principles especially where many players are involved e.g. different departments in an intra-organizational Grid structure or a VO. These players have a great deal of control on the Grid infrastructure and any change or management decision will produce an important effect for all of them. For this reason self-management of the Grid infrastructures and services should apply in terms of how resources will be shared and on what charge so that the participant's incentives remain sound and solid. For example, an internal market mechanism such as a pricing unit ("Grid dollar") should essentially be defined as for the Grid resources to be shared according to well-defined principles and priorities.

New markets

By this criterion we refer to the possibility of the creation of new markets due to the use of Grid technology in existing or in new products, not foreseen before. For example, a company that was selling a software product or service to a customer based on specific commercial licenses (e.g. per machine installation), now can provide another version of the same service over a Grid infrastructure, without the customer having to install the software in his workstation, on a pay-per-use basis where the customer will pay for the times he uses the service only and depending on his requirements such as the QoS needed, the availability of the service, the completion time. The provider will make available different version of his service to accommodate the different requirements of his clients.

Furthermore, and due to the realization and wider adoption of the Grid technology a service provider could offer his product as a number of different stand-alone services (components) which the end-user can utilize together with services from other providers towards a new highly-customized and personalized scalable product or service. All players in this scenario take advantage of the new market foreseen by the realization of Service Oriented Architectures (SOA).

New entry and impact to other markets
As already mentioned, the new market created by the Grid adoption may have a significant impact to other markets too. The fact that applications are now offered on a pay-per-use basis provides new capabilities to SMEs, which can serve as providers to other markets, the barriers of entry to which are thus lower. Indeed, the SME can now develop applications and offer services over virtualised Grid environments (with fewer components, less actual development time and expensive infrastructure owned) and use the computing power of a Grid utility provider in order to offer them to a new market (not achieved before) thus directly impacting and increasing competition of this already established market.

Free-riding
In economics, collective bargaining, and political science, free riders are actors who consume more than their fair share of a resource, or shoulder less than a fair share of the costs of its production. The free rider problem is the question of how to prevent free riding from taking place, or at least limit its negative effects. Because the notion of 'fairness' is controversial, free riding is usually only considered to be an economic "problem" when it leads to the non-production or under-production of a public good or when it leads to the excessive use of a common property resource.

The problem and effects of free-riding are really evident in the context of a Grid virtual organization where resources are shared among and for the common benefit of their participants. A free-rider highly consuming participant limits the common benefit and participates on the expense of other participants. Hence, it is really imperative for internal agreements e.g. SLAs to be implemented among VO-forming participants, the right incentives to be given to prevent free-riding, and penalties to be applied in cases where this is detected.

Information asymmetry, risk and unpredictability-related issues
Information asymmetry arises when one party to a transaction has more or better information than the other party. Typically it is the seller that knows more about the product than the buyer, however, it is possible for the reverse to be true: for the buyer to know more than the seller. Information asymmetry leads to market inefficiency, since not all the market participants do have access to the information they need for their decision-making processes.

In the context of Grid, information asymmetry and issues related to risk and unpredictability arise when participants of a Grid environment (either inter- intra-organizational) have incomplete information about the incentives and repudiation of other participants, such as VO members or internal company departments. This has a negative effect on their willingness to participate in the formation of Grid as well as on their reluctance in sharing their resources and data over the infrastructure. In all cases there is an associated risk and unpredictability of other partners' future behaviour and the origin of their incentives. Apart from the impact of information asymmetry on the sharing of resources by the Grid participants, security issues also impact their willingness to share data especially when sensitive information is to be distributed. This risk also applies to clients. Finally there is an always evident risk of adopting and investing on a new technology especially if this hasn't been fully adopted or if it is based on proprietary implementations.

Performance differentiation and QoS
The objective of performance improvement for application and services constitutes one of the foremost reasons for a company or organization in adopting/moving towards the Grid technology. Thus, it becomes apparent in such cases that the amount of money someone is willing to pay for a service provided over Grid or for such an implementation is strongly dependent on the magnitude of the advantage that this will offer to him in the market. The requirements from the clients/end-users may differ in terms of QoS parameters such as the time of completion, the availability and in this sense it is required to have different and adaptable (but secured by SLA-type agreements) level of services offered by the provider.

5 A Case Study: Analysing and Evaluating the BEinGRID Business Scenarios in Terms of the Identified Economic Issues

Following the identification of the main drives for Grid adoption and having elaborated on the economic issues around them, our next step is to classify the large number of possible Grid Business Scenarios in specific categories to enable us to discuss them further and investigate their impact in real-life scenarios.

In order to accommodate for business examples from different industries we have chosen to analyse the 18 business scenarios from the BEinGRID project (called Business Experiments –BEs in the context of the project). A high-level description of these business cases can be found in [3]. The reasons behind our selection were the following:

- The BEinGRID business cases constitute real-life scenarios in the respect that are implemented by companies that their intention is to enter the Grid market immediately upon the successful completion of the project. Most of these companies did not have any previous experience with Grid Technologies and are currently in the phase of considering Grid adoption by evaluating all the relevant factors both business and technology oriented with emphasis on the former.
- The scenarios cover a range of industries from automotive and film industry to financial and ship building ones and including companies from the whole Grid services provisioning value chain: resource providers, integrators, service providers, end-users etc.

- As members of the BEinGRID consortium we had access to detailed (and sensitive) economic information such as their business models and business plans something that would not be available to us in any other case.

5.1 Classification of Grid Business Scenarios

Firstly, to ease the process of our analysis, the Business Scenarios were classified in three distinct categories, corresponding to the main economic objectives presented and discussed in the previous section. These categories are the following:

- *Category 1: "Grid Business Scenarios with a clear performance-associated benefit"*. This category of scenarios represents those cases that their implementations primarily aim at addressing one of the following limitations: a) additional CPU power needed for executing a demanding application (typical HPC scenario) b) huge amount of data storage/memory is required c) access to heterogeneous, geographically distributed data resources is required.
- *Category 2: "Grid Business Scenarios with a highly collaborative benefit"* i.e. benefit arising from sharing complementary resources among participating organizations. In this case the resulting benefit from Grid adoption comes from sharing data, power and resources utilized for a common scope. Typical examples of this category are intra-organisational Grids and Virtual Organisations and the expected economic benefit in this case could be shared among all participants in contrast to the first category where the main economic benefit is anticipated from the end-user where the service or application will be provided. Also, the services of this category cannot be provided by a single provider since data or other resources are necessary to be obtained from other providers.
- *Category 3: "Grid Business Scenarios exploiting the component-based soft-ware paradigm"*. This category comprises those business scenarios involving a service provider that offers applications on a pay-per-use basis rather than by means of licensing or long term static agreements and thus exploiting to the most the concepts of the next generation Service Oriented Architectures (SOA).

The classification of each of the BE's was based on analysing the technical context, business motivation and detailed work planned for the BE, as this was described in the relevant BEinGRID technical documents. In cases where a BE belonged to more than one categories, the decision was based on the prioritisation of the BE objectives as this was presented in the internal BE description and in some cases based on the feedback provided by us after contacting and interviewing the BE leading partner.

Our preliminary analysis with regard to the business scenarios and by examining their initial business plans provided to us in the context of the project, indicated that approximately 70% of the cases belonged to Category 1, 25% in Category 2 and only 5% in Category 1.

5.2 Impact of the Economic Issues in the Business Scenarios

Following the identification of the most important economic issues applicable to the Grid computing adoption in Section 4, the business scenarios were analysed in terms of these issues in order to investigate their relevance to the specific cases, the extent that these apply and therefore the importance that should be given to those by the partners involved in these experiments.

The BEs were evaluated using 3 different grades based on the applicability of each economic issue. The three grades were the following:

Grade A – Strong Impact: Economic issues of this kind exist in this business case; their impact is very strong and should be carefully addressed.
Grade B – Average Impact: Economic issues of this kind may exist depending on the scenario configuration, or may exist in the future, their impact is and therefore should be analysed.
Grade C – Weak Impact: Economic issues of this kind do not exist or exist but their impact is considered weak and thus it is not vital to be considered at this point.

Inputs for our evaluation were provided by the partners of the business experiments in terms of their business models and plans, technical descriptions of their scenarios and by personal interviews. The results of the evaluation for the experiments are presented in a tabled form in Appendix A.

5.3 Discussion on the Impact of Specific Economic Issues in the Business Scenarios

Our evaluation of the economic issues identified earlier with respect to the specific business scenarios resulted in a number of observations per economic issue examined. Due to space constraints only two of them are listed below as examples.

Network externalities
Network externalities are very evident in many of the experiments that involve the forming of a virtual organization to serve a common scope such as the execution of a complex simulation. The gained benefit for each organization is proportional to the number of organizations participating and offering their resources for the common purpose. For example in "BE02: Business workflow decision making" in order for the risk simulations for the film production industry to be as comprehensive and sound as possible, information must be collected from many of the involved actors: film editors, special effects producers, animators etc – the more obviously the better. If the information is limited then the benefit for the end-user, i.e. the quality of risk-related results given to the producer, becomes questionable, thus decreasing the willingness of the producer in participating in such a virtual organization. The same characteristics can be found in BE10: Collaborative environment in the supply chain management where the number of participants increases the total benefit and vice versa, thus influencing the amount a potential customer is willing to pay for the same service. These observations are in line with our Category 2: "Grid Business Scenarios with a high collaboration benefit" economic characteristics discussed previously.

Information asymmetry, risk and unpredictability-related issues

As discussed in the previous section, information asymmetry and issues related to risk and unpredictability arise when participants of a Grid environment possess incomplete information about the incentives and repudiation of other participants, such as VO members or internal company departments. This has a negative effect in their willingness to participate in the Grid environment and in their reluctance for sharing their resources and data over the infrastructure. In these cases there is also an associated risk and unpredictability of the new partners' behaviour. These issues are more evident in the studied Grid scenarios where the Grid participants are not well-known before and depending on their numbers i.e. in the more "open/loose to participation" cases of Grid structures. On the other hand, more "closed" type of Grids, such the virtual organisations formed by a company's departments (enterprise Grids), are obviously less susceptible to such issues. Examples of the former are "BE10: Collaborative environment in the supply chain management" and "BE14: New product and process development" whereas of the later is "BE12: Sales management system".

For example, in BE14 let's consider a small firm that intends to run a complex CAD simulation for a potential new product. They have tried to run this simulation on their few workstations but couldn't complete it due to the insufficient power available from their machines. Using Grid technology i.e. "renting" infrastructure from a provider seems as an attractive option to them instead of buying new PCs or a new better CAD tool. However, their lack of expertise in computing and the fact that this is a new product does not enable them to estimate exactly the amount of CPU power and memory that will be needed from their CAD tool in order to perform these simulations. On the other hand, the computer experts from the Grid provider side, having used CAD tools extensively in the past and having rented their infrastructure to other companies for the same purpose in the past are in a better estimate the power needed for their simulation. If this information is not disclosed to the buyer (the small firm) could create a situation where they will pay to utilise more resources (to be on the safe side) than those actually needed for their product thus causing a market inefficiency.

6 Conclusion and Further Work

Grid technology has the potential to revolutionise the way services are distributed and executed over heterogeneous dispersed infrastructures in the future. Lessons learnt from recent past have taught us that technological maturity stand-alone cannot drive a new technology forward. Business and economical drivers should be considered as equally important. Along that direction, in this paper we have tried to identify and analyse a number of dominant economic issues that could act as both acceptance drivers as well as impediments and therefore should taken into account by industrial actors considering the adoption of the Grid for their business. These issues include the associated economies of scale/scope, information asymmetry, self-management issues, network externalities, free-riding, impact to new markets etc. We examined these in the context of a case study with real-life scenarios. Furthermore, we evaluated them in terms of their impact/influence in the decision process of whether a company should adopt the grid or not in the scenarios under consideration. Further work and analysis will include specific proposals on tackling these issues to be applied in an array of different industries. Finally, further work will include the definition of a decision model and associated methodology to be utilised by both Grid experts and business people for deciding towards the Grid adoption, based on the factors presented in Section 4.

Acknowledgements

The authors wish to thank T. Papaioannou and S. Routzounis for their contribution to the material of Section 2.

References

1. Forge, S., Blackman, C: Commercial Exploitation of Grid Technologies and Services, Drivers and Bariers, Business Models and Impacts of Using Free and open Source Licensing Schemes, SCF Accosiates for DG Information Society and Media (2006)
2. http://cordis.europa.eu/ist/grids/index.html
3. http://www.beinGrid.eu
4. http://cordis.europa.eu/fp6/
5. Foster, I., Kesselman, C., Nick, J.M., Tuecke, S.: The Physiology of the Grid. An Open Grid Services Architecture for Distributed Systems Integration. Open Grid Service Infrastructure WG, Global Grid Forum (2002)
6. http://www.globus.org/ogsa/
7. Cheliotis, G., Miller, S., Woodward, J., OH, D.: Questions for Getting Smarter on Creating a Grid Market Hub, Grid MarketPlace RoundTable. GECON'06, Singapore (May 2006)
8. Kenyon, C., Cheliotis, G.: Grid Resource Commercialization. International Series in Operations Research and Management Science, Netherlands, ISSU 64, 465–478 (2003)
9. Cheliotis, G., Kenyon, C.: Autonomic Economics: Why Self-Managed e-Business Systems Will Talk Money. In: IEEE Conference on E-Commerce '03 (2003)
10. Gray, J.: Distributed Computing Economics. Microsoft Research Technical Report: MSR-TR-2003-24 (also presented in Microsoft VC Summit 2004, Silicon Valey, April 2004) (March 2003)
11. http:// www.Gridbus.org/
12. Buyya, R., Abramson, D., Venugopal, S.: The Grid economy. In: Special Issue of the Proceedings of the IEEE on Grid Computing, IEEE Press, Los Alamitos (2005)
13. Brunelle, J., Hurst, P., Huth, J., Kang, L., Ng, C., Parkes, D.C., Seltzer, M., Shank, J., Youssef, S.: Egg: an extensible and economics-inspired Open Grid computing platform, GECON (2006)
14. http://mnetproject.org/
15. Weng, C., Li, M., Lu, X., Deng, Q.: Economic Based Resource Management Framework. CCGrid'05 (2005)
16. MacKie-Mason, J.K., Osepayshvili, A., Reeves, D.M., Wellman, M.P.: Price Prediction Strategies for Market-Based Scheduling. In: Proc. of ICAPS' 2004 (2004)

Appendix A: Evaluation of the Impact of the Economic Issues in the BEs

Application	Economies of Scale	Economies of Scope	Extern alities	Self-manag ement	New market s	Impact to other markets	Free-riding	Info. Asym metry	Perform. Differenti ation
Computational Fluid Dynamics & Computer Aid Design	A	B	C	A	A	B	A	B	B
Business workflow decision making	A	C	A	A	A	B	B	B	A
Visualization & virtual reality	A	B	C	A	A	B	B	C	B
Financial Portfolio Management	A	A	C	A	A	A	B	A	A
Retail Management	A	C	C	A	B	B	A	B	A
Groundwater modelling	A	C	C	A	A	C	B	A	B
Earth Observation	A	A	B	A	A	B	A	B	B
Engineering and business processes in metal forming	A	A	A	A	A	B	A	B	A
Distributed online gaming	B	A	A	B	A	A	A	B	A
Collaborative environment in the supply chain management	A	B	A	A	B	B	A	A	B
Risk management	A	C	B	A	A	B	B	A	A
Sales management system	A	C	B	A	B	C	B	C	A
Textile Grid portal	A	A	A	A	A	A	A	A	B
New product & process development	A	B	B	A	A	B	A	A	A
Virtual engineering workplace for financial e-services	A	A	B	A	A	B	A	B	B
Ship building	A	C	C	A	B	B	A	B	A
Logistics & Distribution	A	C	B	A	A	C	B	C	A
Seismic imaging & reservoir simulation	B	A	A	A	A	B	A	A	B

Decision Factors of Enterprises for Adopting Grid Computing

Junseok Hwang and Jihyoun Park

Technology Management, Economics and Policy Program,
Seoul National University, Korea
{junhwang,april3}@snu.ac.kr

Abstract. Enterprises adopt the autonomous Grid computing as their ICT infrastructures in order to improve business flexibility and scalability. In this technology adoption process, Grid solution providers take roles as technology deciders and distributors. This paper investigates the effective decision factors that influence enterprises' technology choices through a survey among Grid solution providers. The survey evaluates the relative importance of decision factors using the conjoint analysis method. Survey results show that the interoperability is the most indispensable element to successfully install the Grid in enterprise ICT infrastructures. Based on the findings, this paper explains the market context that impacts the revealed technology choices.

Keywords: Grid for enterprises, Conjoint analysis, Technology choice.

1 Introduction

Enterprise ICT infrastructures have evolved to extend scalability and flexibility in utilizing the computing resources. The business platform has transformed from a mainframe system to a client-server system and eventually to today's web based system [3]. The IT outsourcing trend in 1990's clearly shows that enterprise systems need efficient and advanced IT operations and management for supporting their core businesses [2]. Enterprise IT professionals are continuously looking for more capacity and adaptability to respond to a rapidly changing business environment. Grid computing, which was initiated from solving large-scale mathematical and scientific problems, is a possible alternative that satisfies such needs. It also builds up a new computing environment for businesses. Merged with web services and service-oriented architectures, Grid computing architecture pursues on-demand deployment of resources based on exactly what consumers need in terms of both quantity and quality over distributed environments. This new computing environment for enterprises is called utility computing.

Grid computing is an IT infrastructure sharing heterogeneous resources beyond administrative boundaries. Openness is its major characteristic. Ideally, any computing resource connected to a Grid network is utilized without knowing the source. As Grid computing advanced into commercial areas, how to apply the openness to the business became an issue. While business continuity is critical in

D.J. Veit and J. Altmann (Eds.): GECON 2007, LNCS 4685, pp. 16–28, 2007.

enterprise ICT infrastructures, the openness of Grid is accompanied with uncertainty problems in controlling resource availability, system failures and security. Moreover, today's technologies are not mature enough to overcome those limits. The benefit of openness for enterprise businesses, however, cannot be negligible because that would allow any kind of necessary resources available for enterprises to develop competitive business models (BMs).

This paper investigates these trade-off problems that enterprise IT experts face when they plan Grid computing as their enterprise ICT infrastructure. IT experts in an enterprise select a range and level of a system's openness and business flexibility according to their judgments and preferences among other various decision factors. This paper uses a conjoint survey technique to analyze the relative importance among Grid attributes for enterprise ICT infrastructures.

The paper is organized as follows: in section 2, technical backgrounds are described, especially in the controversial points of Grid computing as an enterprise ICT infrastructure. Section 3 explains the empirical research procedure. Section 4 defines who leads the technology trends and what their decision factors are. Survey techniques to measure the value of each decision factor are described in section 5. Theoretical backgrounds to interpret the survey results are explained in section 6. Section 7 shows the results of an empirical research carried out among Korean Grid solution providers. In section 8, conclusions and policy implications are provided.

2 Backgrounds

The needs for agility and scalability in enterprise ICT infrastructures are ever increasing in rapidly changing business environments [2]. These requirements can be resolved by adding more resources into the system but this causes a management burden, especially if the company's core business is not related to IT. Another alternative is using external resources distributed over networks [4, 5, 7]. The concept of Grid technology, which is resource sharing across industry boundaries, can be used to build reliable, scalable and secure distributed systems for enterprises in response to these new challenges [4,5]. The Grid environment has another benefit for enterprise business. Combining with web services, the Grid service enables out-of-box implementation of enterprise infrastructures and accelerates the speed of development of business models so as to increase the company's competency in the market.

Although enterprise IT experts prefer to have a flexible infrastructure, there exist other decision factors that they must take into consideration. If an enterprise ICT infrastructure is in charge of the company's core businesses, the primary goal of the system is to maintain business continuity. In this case, the average availability required for the enterprise server is over 99%. An ideal Grid, however, does not necessarily guarantee such a level of resource availability. The Grid includes three main features: heterogeneity, scalability and dynamicity or adaptability [6]. Due to the last feature, the probability of a system failure is increased while the overall system can achieve the maximum utilization of available resources.

Grid offerings are classified in many ways. The level of geometrical distribution and hardware heterogeneity covered by the offering are one of the possible criteria. According to these criteria the Grid is differentiated into departmental Grids, enterprise Grids, partner Grids, and open Grids [1]. The current status of Grid computing for the enterprise system that the market understands still stays at the level of partner Grids [1]. The partner Grid utilizes resources only within trusted organizations and has limits in functionality and capacity extension. As technology advances, the Grid system for enterprises is evolving toward open Grids extending the market. The properties of partner Grids and open Grids are described in Table 1.

Table 1. Partner Grid versus Open Grid

	Partner Grids	Open Grids
Resource	Resources inside trusted organizations	Any resources connected to Grid networks
Application	Specific applications enabled for the Grid environment	Any type of applications, which follows the OGSA
Interoperability between Grid solutions	Possible but mostly proprietary solutions	Perfect interoperability
Strength	Guaranteed quality of services (QoS), trusted environment	Strong scalability, easiness in developing new services
Weakness	Limited capacity	Security problem, QoS issues

An open Grid has two aspects: open standard and open resources. If the current Grid for enterprise ICT infrastructures is diagnosed from this view, it can be said that the solution follows the open standard but utilizes only the limited resources inside the trusted environment.

The open standard in the Grid system is the foundation establishing perfect interoperability, extensibility, portability and sharing relationships among participants [4]. The open standard specification, the Open Grid Services Architecture (OGSA), was first introduced at the forth Global Grid Forum (GGF) in January 2002. The OGSA offers a high-level standard architecture. Currently, it is not yet specified to implement programming codes [5, 7]. Therefore, it is possible to have different solutions in implementing a Grid middleware.

The second criterion of open Grids is the resources of open environment. Using open resources provided in public resource pools have several advantages. A company can obtain a variety of hardware and software resources, optimize the utilization, save costs and quickly respond to exceptional events. However technical limitations and psychological resistances prevent enterprises from adopting open environments. Enterprise IT experts are afraid that infrastructure sharing may weaken the system's security level and increase uncertainties in quality in the case of network congestion.

3 Methodology

An individual enterprise faces a choice problem among technological alternatives of the Grid as seen in section 2. In this context, the main research questions are as follows: (1) what are the key factors for adopting Grid computing, (2) how much is each factor important compared to the others.

To solve the problem, this paper first defines who leads the technology trend. These technology leaders are the targets of the research. Second, the technology leaders' decision factors are defined. The conjoint analysis survey technique is used to investigate relative importance among key decision factors and to see how much the openness of the Grid in terms of protocols and resources has influenced enterprise ICT infrastructure. Two empirical models are designed basing on the random utility model. The first model captures the Grid professionals' preferences among the market share of the open Grid system, the open standard, the uncertainty in QoS, the complexity in business model development and the ease in implementation. In addition, the second model captures the difference by the type of company that affects technology choices. The empirical study was conducted by the 41 Grid professionals in Korea in 2006.

4 Discussion of Decision Factors

4.1 Technology Adoption

In the initial stage of a new technology development and adoption, preferences of technology solution providers are rather influential to the process than those of solution adopters. When the target technology is first introduced in the market, people are likely to depend on professional analysis for that technology. Fig. 1 presents a situation that strategic analysis and investments by solution providers derive market demands when the market has little experience with the technology. These solution providers are also opinion leaders in the market. In product-marketing studies, the word-of-mouth effects and the role of opinion leaders are considered as key factors in impacting brand promotion [11]. Thus, to identify critical factors of Grid for enterprises, research will be conducted by analyzing the view of the Grid solution providers rather than the Grid solution adopters because Grid solution providers lead and affect the decision of adopters.

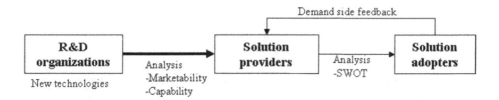

Fig. 1. A model for demand analysis of enterprise IT technology

4.2 Decision Factors

Section 2 explains several trade-offs that should be taken into consideration when Grid computing is adopted for an enterprise infrastructure. Five selected trade-off factors, which affect the decision of the adopting level and the scope of Grid technologies, are summarized in Table 2 with the expected influence sign, positive or negative, of each factor toward open Grids.

Table 2. Trade-Offs among Grid System Technologies

Decision factors	Expected influence to open Grids
Market share of the open Grid system	+
Open standard	+
Uncertainty in QoS	-
Complexity in business model development	-
Easiness in implementation	+

An enterprise can choose a desired level of Grid technology between partner Grids and open Grids. If an enterprise considers the future enterprise ICT infrastructure is an unprecedented dynamic computing environment, it will select open Grids in order to survive in fierce competition. The enterprise can get huge advantage over business development and infrastructure implementation by choosing the open Grid technology. However it needs to accept the system that may cause uncertainty problems in performance. For enterprises that prefer the open Grid, compliance to the open standard is essential to making the open environment.

On the other hand, an enterprise, which is cautious of confidential information exposure and requires a high level of system availability, may not want to use the open Grid. The tendency to observe the open standard is weaker. Although the academic definition of the Grid proposed by Ian Foster is characterized as decentralized, open protocols and nontrivial QoS [12], commercial systems are interested in tangible, operational and profitable technologies [1].

The current market share of the open Grid system also affects enterprises' technology choice. The network industry is distinguished by having strong network externalities [13]. Network externality means that the utility derived from the consumption of a good is affected by the number of other people using similar or compatible products. Similarly, the value of open Grids increases dramatically as the amount of software diversity and hardware capacity increases.

5 Survey

To analyze decision factors by calculating individual demand for each attribute of the Grid system for enterprises, this paper adopts a conjoint analysis methodology. The conjoint analysis is a survey technique that estimates the amount of demand and relative importance among attributes of a product or service. As a stated-preference survey methodology, it is most widely used for modeling the trade-offs and decision processes. In the survey, several hypothetical alternatives of a product are provided to respondents with different levels of the product's attributes. This procedure simulates a situation that consumers face when they purchase products or services in a market [8].

Batt and Kats evaluated the influence power of six attributes of the enhanced voice mail in the USA in early 1993 using the conjoint analysis. They derived a demand and revenue model and predicted the demand and revenue with various types and price scenarios of the enhanced voice mail through market simulations [8]. Schoder and Haenlein studied the relative importance among attributes that affect a seller to construct trust with a potential buyer in online commerce using the conjoint analysis.

Their result gave special support for building a legislative system in online commerce [10]. A combined methodology, which incorporates a diffusion model to the conjoint analysis, is actively applied for analyzing how competing products with different technological attributes divide a market and are diffused throughout the market. Jongsu Lee et al., for example, measured the demand for the large screen TV market in Korea. The study forecasted demand for each competing product – projection, LCD and PDP TV. The conjoint analysis was used for capturing consumers' preferences for each of the different TV types [9].

5.1 Conjoint Analysis Survey Technique

In section 4.2, five attributes to describe a Grid system technology with its market environment was discussed. Table 3 shows the five decision factors and experimentally designed levels for each factor.

Table 3. Properties of the Grid System and Market Environment for Composing Conjoint Cards

Property	Description	Level
Market share of the open Grid system	An assumed market share of open Grids over the whole Grid systems for enterprises.	25% 50% 75%
Open standard	Whether or not the Grid system is complied with the OGSA standard.	Yes No
Uncertainty in QoS	A probability of system failure in the Grid system due to data hacking, virus and other resource problems. The probability is raised higher as resources in open environments are used.	0% 1% 2% 3%
Complexity in business model development	A period to develop a new service using the Grid system from when customers' demand occurred.	3 months 6 months 12 months
Easiness in Implementation	An implementation method by which the Grid system is installed on a customer's site.	Customization Out of box

The first attribute is the market share ratio of the open Grid system. It has three different values, 25%, 50% and 75% that represent the portion of the market belonging to the open Grid system. A greater market share implies that the open Grid system is dominating the market. This attribute estimates how Grid system developers respond to various market situations. The other 4 attributes are different properties of the Grid system that can be selected by the respondents. The open standard property represents whether or not the virtual Grid system follows the open standard or OGSA. For the uncertainty in QoS, a range of 0% to 3% system failure probability is used for representing the increasing characteristics of it as computing resources in open environments are more involved. The complexity in developing business models is measured by the development period of a new service in the assumed Grid system. It also explains how quickly the enterprise system is able to respond to external stimuli. The last attribute, the ease in implementation of the Grid system is especially important for the solution providers because the easy, fast implementation is one of the keys for development companies' competency. It contains two levels: one is the customization method that can satisfy more specific needs for each customer while it takes more time and finances. The other is the out-of-box method that makes the installation project short and simple.

Five attributes with 2 to 4 levels yield 144 cases in total. Using the factional factorial design (FFD), the number of alternatives is reduced to 16. These cases are grouped into 4 sets and provided to respondents in the form of Table 4. Respondents are asked to rank the provided four alternatives from the most favorite hypothetical bundles of attributes to the least. Generally there are three ways that respondents can indicate their preferences. People can select only the best alternative, rank all the alternatives, or rate the value of each alternative respectively. The rank ordered approach was adopted because it covers more information from customers and it is suitable for representing the ordinal preference [9].

Table 4. Example of one set of Conjoint cards

1		2		3		4	
Open Grid market share	25%	Open Grid market share	75%	Open Grid market share	25%	Open Grid market share	50%
Open standard	N	Open standard	Y	Open standard	N	Open standard	Y
Uncertainty	1%	Uncertainty	3%	Uncertainty	2%	Uncertainty	3%
BM development	12M	BM development	3M	BM development	3M	BM development	12M
Implementation	Out of box	Implementation	Out of box	Implementation	Customization	Implementation	Customization
( )		( )		( )		( )	

5.2 Data

The survey was conducted by the 41 Grid professionals working at member companies of the Grid business association (GBA) in Korea from the 21st of November 2006 to the 14th of December 2006. They are system engineers, software programmers, and technical sales engineers who have at least one-year experience with the Grid. The survey was performed through emails, telephone interviews and face-to-face interviews. 38 questionnaires were obtained as valid data. Among them 17 were from people who worked for general IT solution companies and the rest were from people who belonged to Grid specialized solution companies.

6 Analysis Models

To investigate how given decision factors influence a user's choice, a random utility model is used. This method establishes an equation which measures how much each factor changes the user's utility level. By the model, the utility function U_{ij} that the ith individual selects the jth alternative among J choices is defined as follows [18]:

$$U_{ij} = V_{ij} + \varepsilon_{ij} = x_j\beta + \varepsilon_{ij} \; . \tag{1}$$

U_{ij} is composed of the deterministic utility, V_{ij} and the stochastic utility, ε_{ij}. The deterministic utility is derived from observable data whereas the stochastic utility is unobservable and estimated as a random disturbance. The deterministic utility is decomposed into x_{ij} and β, where x_{ij} is the vector of attributes associated with the jth alternative and β is the vector of the coefficients of attribute vector. The stochastic utility ε_{ij} is assumed independently and identically distributed (iid) according to type-I extreme value distribution.

In the empirical model of this research paper, U_{ij} is the utility of the ith Grid solution provider when the jth Grid system is chosen to be developed. With the empirical specifications defined in section 5.1, (1) is re-written as follows:

$$U_{ij} = \beta_{iMA}MA_{ij} + \beta_{iST}ST_{ij} + \beta_{iUN}UN_{ij} + \beta_{iBM}BM_{ij} + \beta_{iIM}IM_{ij} + \varepsilon_{ij} \; . \tag{2}$$

MA represents the market share of the open Grid system over the whole market of the Grid system for the enterprise. ST is a dummy variable, which takes the value of 1 if the selected Grid system follows the OGSA standard. UN means the uncertainty probability in the selected system's resource trustworthiness. BM is a variable which evaluates the degree of flexibility in business environment supported by the selected Grid system. The value is represented by the development period of a new business model. As the period is longer, the business environment is more rigid and complex. IM is another dummy variable which indicates 1 if the installation of the selected Grid system is carried out in the way of out-of-box.

The previous utility function modeling is based on an assumption that marginal utility on each attribute is the same for the entire respondent. Thus the first model includes no specific terms reflecting heterogeneity of individuals. Otherwise, we can assume that respondents, who are Grid professionals, show different preferences according to the market strategies that their companies have. Here, the second model is proposed reflecting the solution provider's characteristics.

$$U_{ij} = \beta_{iMA}MA_{ij} + \beta_{iST}ST_{ij} + \beta_{iUN}UN_{ij} + \beta_{iBM}BM_{ij} + \beta_{iIM}IM_{ij} + \beta_{iCO}CO_{ij} \times ST_{ij} + \varepsilon_{ij} \; . \tag{3}$$

The interaction term $CO_{ij} \times ST_{ij}$ is added to the former model. This term proposes that the strategy toward the open standard is assumed to be different between Grid-specialized development companies and general IT solution companies. CO is a dummy variable, which represents type of a company that a respondent belongs to. This variable is 1 if the respondent works for a Grid specialized company and 0 if he or she belongs to a general IT company. ST is, as described before, a dummy variable which indicates whether or not the company's Grid system observes the open standard. For convenience, the former model is hereafter named Model 1 and the latter is Model 2.

The coefficient of each term is the influence power of the decision factor. As the conjoint survey of this research uses the rank-ordered approach, a rank-ordered logit model is usually used for the estimation of coefficients [14, 15]. Coefficients are calculated by applying the maximum likelihood estimation method on data obtained from the survey and the maximum likelihood function is derived from the choice probability of the rank-ordered logit model [18].

7 Results and Analysis

7.1 Effective Factors

The estimated coefficients are provided in Table 5. For both models, coefficients of 4 variables, which are directly related to the technology itself, are statistically significant in a significance level of 1% while the coefficient of the market share variable of the open Grid is estimated insignificant.

The value of the open Grid market share variable is even almost 0. Contradictorily, the coefficient of the open standard compliance variable has a strongly positive value, 0.633. This indicates that the two variables show huge differences in interpretation, although they seem to convey similar meanings about interoperability characteristics of the Grid. Solution providers do not pay much attention to the current market share of open Grids, not because the Grid does not express the network externality, but just because they have strong tendency to obey the open standard. This result is also related to their expectations about future technology trends. If solution providers expect the standardization procedure will be finished within short time, it is better for them to join the specification rather than to develop other systems. The bandwagon effect is much stronger in network industries than in any other industry areas [13].

The negative sign of the uncertainty in QoS and the complexity in business model development is intuitive. Grid technology will be less diffused in the enterprise ICT infrastructures as the uncertainty and complexity in the constructed computing environment is greater. The positive value of the ease in implementation variable means the out-of-box style of system installation and management is preferred. This also reflects that people favor business flexibility and agility. Again, it is a consistent result with the negative coefficient of the complexity in a business model development.

In the meantime, the interaction term that measures the different attitude toward the open standard according to types of a solution provider company is estimated as statistically insignificant. The fact that the value is negative can be misinterpreted so that the Grid specialized companies may want to develop their own proprietary products for surviving in the fierce competition of the market. However the negative value of this variable can be better explained by the misunderstanding of the term 'open standard' among the respondents. 'Open standard' is often mixed up with the 'Globus open source project', while in fact an individual company can incorporate its own creative parts into the product and simultaneously follow the OGSA standard. The leading companies such as United Device, DataSynapse and Platform Computing also pursue this product differentiation strategy supporting perfect interoperability with the OGSA. It is not a sustainable strategy for any company to deviate the standard in network industries where the rule of 'winner takes all' strongly governs.

Table 5. Coefficient Estimates from Conjoint Analysis

Variable	Description	Coefficient (t-value)	
		Model 1	Model 2
	Observations	152	152
MA	Market share of the open Grid system	-0.003	-0.003
		(-1.246)	(-1.265)
ST	Open standard	0.633	0.798
		(5.216)**	(5.079)**
UN	Uncertainty in QoS	-0.263	-0.267
		(-5.187)**	(-5.255)**
BM	Complexity in business model development	-0.118	-0.119
		(-7.796)**	(-7.854)**
IM	Easiness in Implementation	0.320	0.321
		(2.633)**	(2.639)**
CO*ST	Grid specialized companies' preference toward		-0.284
	the standard		(-1.616)

7.2 Relative Importance

Using the estimated coefficient of each attribute, the relative importance among attributes is calculated and provided in Table 6. The procedure to obtain the relative importance is first to calculate the difference between the highest value and the lowest value for each attribute and then compare the result with other attributes.

The result reveals that the open standard compliance is the most important decision factor than any other attribute for Grid solution providers when they select the level and scope of the Grid technology. The relative importance of the open standard compliance reaches 47.34%, almost half, therefore, influencing a decision. If the effect of the Grid specialized companies is excluded, the weight of the open standard is even more greatly increased as shown in model 2. The next influential factor is the ease in implementation of which the relative importance is 23.94%. The uncertainty probability in system QoS has a 19.67% relative importance. The relative importance of the complexity in the business model development variable, which represents the degree of flexibility in business supporting environments, is relatively low. Only about 9% are considered in the decision process of technology choices. It is worth mentioning that this probably resulted from the fact that the research target of this survey is not solution adopters but solution providers. The ease in implementation, which represents business flexibility and agility on the solution providers' side, illustrates about 3 times more impact than the previous one on the decision process. The impact of the market share of the open Grid system can be ignored.

Although the result reports the solution providing companies have strong wills to follow the open standard, it does not guarantee their support on the open Grids. That is because the open Grid system includes not only the open standard but also the share of open resources as explained in section 2. Particularly, the latter condition prevents the open Grid from spreading out in the commercial area. The results of this research reconfirm that the openness of Grid in terms of protocol has positive influence over enterprise ICT infrastructure, although in terms of resources the influence is negative.

Table 6. Relative Importance among Grid System Attributes

Variables	Descriptions	Model 1		Model 2	
		Part worth	Relative importance (%)	Part worth	Relative importance (%)
MA	Market share of the open Grid system	-0.003	0.22%	-0.003	0.20%
ST	Open standard	0.633	47.34%	0.798	52.92%
UN	Uncertainty in QoS	-0.263	19.67%	-0.267	17.71%
BM	Complexity in BM development	-0.118	8.83%	-0.119	7.89%
IM	Easiness in Implementation	0.320	23.94%	0.321	21.28%

8   Conclusions and Policy Implications

The enterprise ICT infrastructure is evolving to the utility based open computing environment in order to enhance flexibility and scalability in business. The open computing environment for enterprise systems is accomplished by both following the open standard (OGSA) and using resources provided by the shared environment. Even though the trust issues in open and shared computing environment have not yet been resolved, this trend will be accelerated as the technology advances and the requirement for flexibility increases.

This paper finds that the interoperability is the most indispensable element to successfully install the Grid in enterprise ICT infrastructures. Common interfaces and protocols should be carefully followed throughout the entire system regardless of whether or not the system currently needs communication and cooperation with external systems. However, the open computing environment does not necessarily require one unified middleware solution. The only requirement is an open and agreed standard specification, OGSA. Strategies toward how and when each attribute of the open Grid system is achieved may differ from vendor to vendor based on their analysis on the technology and market. Therefore if policy makers enforce a certain type of middleware framework to apply for all the related Grid products by constraint, it rather results in inhibiting the technology advancement and free competition in the market even though the intention aims to enlarge the open computing environment. The proper role of policy makers or related associations is to aid R&D of the open Grid system in order to resolve its trust problems, and this will be followed by a natural market expansion.

The contribution of this paper is distinguished by three points. First, this study gathered empirical data about the major technological and market environmental variables of the Grid for enterprise ICT infrastructures. We quantified their relative importance and positive/negative influential power with the econometrical methodology. Second, this paper separated two characteristics of the open Grid system. One is the open standard and the other is the shared resource from open environments. Under this framework, it is shown that even though the entire market participants agree to the major premise of the open Grid system, discrepancies among Grid solution vendors may exist when doing detailed actions such as participating efforts or timing for a proposed policy. Third, this paper targeted technology professionals rather than its consumers to forecast the technology demand.

Considering that the technology adoption for the enterprise ICT infrastructures is generally conducted through the professional consulting, this approach strengthens the practicability and the reality of the analysis.

This study has several limitations, nevertheless. First, the conjoint analysis was conducted only for the solution providers. As the technology is settled and experiences with it are accumulated in the market, the demand side feedback that is customers' preferences on technology attributes becomes more and more important. When this is so, additional conjoint analysis for consumers is needed for examining the validity of this research. Second, the empirical data are gathered only within Korea. The results and policy implications of this paper cannot be simply generalized and applied to other circumstances. Lastly it is certain that the Grid will change organizational structures and business environments of enterprises by introducing new methods of IT infrastructure management and pricing. The analysis, however, does not reflect enterprises' reaction and concern about those changes because they are hardly quantifiable. Further studies are required to solve these issues so that enterprises can have the comprehensive and balanced perspective on both technological and organizational aspects of the Grid.

Acknowledgments. We would like to thank Dr. Jongsu Lee for suggestions and comments on the early discussion of this paper, and Dr. Jorn Altmann for detailed review and advice.

References

1. Plaszczak, P., Wellner Jr., R.: Grid Computing The savvy manager's guide. Morgan Kaufmann Publishers, San Francisco (2005)
2. Boden, T.: The Grid enterprise – structuring the agile business of the future, BT Technology Journal, 22(1) (January 2004)
3. Lynne, M.: Markus and Cornelis Tanis, The enterprise system experience- from adoption to success, Framing the domains of IT management: projecting the future through the past, Pinnaflex Educational Resources, Inc. Cincinatti (2000)
4. Foster, I., Kesselman, C., Tuecke, S.: The anatomy of the Grid: Enabling Scalable Virtual Organizations, International Journal of High Performance Computing Applications (2001)
5. Foster, I., Kesselman, C., Nick, J.M., Tuecke, S.: Grid services for distributed system integration, Computer (June 2002)
6. Baker, M., Buyya, R., Laforenza, D.: Grids and Grid technologies for wide-area distributed computing, Software Practice and Experience (2002)
7. Strong, P.: Enterprise Grid Computing, Queue (July/August 2005)
8. Batt, C.E, Katz, J.E: A conjoint model of enhanced voice mail services. Telecommunications Policy 21(8), 743–760 (1997)
9. Lee, J., Cho, Y., Lee, J.D., Lee, C.Y.: Forecasting future demand for large-screen television sets using conjoint analysis with diffusion model, Technological Forecasting and Social Change (2006)
10. Schoder, D., Heanlein, M.: The relative importance of different trust constructs for sellers in the Online word, Electronic Markets, 14 (2004)
11. Lilien, G.L., Rao, A.G., Kalish, S.: Bayesian estimation and control of detailing effort in a repeat purchase diffusion environment, Management science, 27(5) (May 1981)

12. Foster, I.: What is the Grid: a three-point checklist, GridToday (July 2002)
13. Shy, O.: The economics of network industries. Cambridge University Press, Cambridge (2001)
14. Calfee, J., Winston, C., Stempski, R.: Econometric issues in estimating consumer preferences from stated preference data: a case study of the value of automobile travel time, The review of economics and statistics (2001)
15. Train, K.: Discrete Choice Methods with Simulations. Cambridge University Press, Cambridge (2002)
16. Mahajan, V., Muller, E., Bass, F.M.: New product diffusion models in marketing: A review and directions for research, Journal of Marketing, 54 (January 1990)
17. Shapiro, C., Varian, H.R.: Information Rules, Harvard Business School (1998)
18. Train, K.E.: Discrete Choice Methods with Simulation. Cambridge University Press, Cambridge (2003)

Taxonomy of Grid Business Models

Jörn Altmann[1,2], Mihaela Ion[1], and Ashraf Adel Bany Mohammed[2]

[1] Intl. University, Bruchsal, Germany
[2] Seoul National University, Seoul, South-Korea
jorn.altmann@acm.org, mihaela.ion@i-u.de, ashraf@tepp.snu.ac.kr

Abstract. Grid Computing, initially intended to provide access to computational resources for high-performance computing applications, broadened its focus by addressing computational needs of enterprises. It became concerned with coordinating the on-demand, usage-based allocation of resources in dynamic, multi-institutional virtual organizations, and eventually creating new business models based on this technology. This trend in Grid computing holds a lot of potential in many industries with respect to saving costs, improving efficiency, creating new services and products, increasing product quality, as well as improving collaboration between companies. This will change the way business is done and it will change our classical view of the value chains, its stakeholders, and their roles. However, in order to encourage more companies to adopt Grid computing, value chains have to be explained and business models have to be understood. This paper makes a first move in this direction. It analyses existing business models. Based on the result of the analysis, it formally defines a taxonomy of existing and future roles that a stakeholder can take on within the value chains of the Grid and gives examples of those roles. Finally, this paper applies the taxonomy to two reference business models: utility computing and software-as-a-service.

Keywords: Grid Computing, Grid Economics, Business Models, Functional Roles, Taxonomy, Utility Computing, and Software-as-a-Service (SaaS).

1 Introduction

Grid Computing started out of the necessity to solve computational-intensive scientific problems that needed more resources than available at a single high-performance computing center (HPCC). Using Grid technology, storage capacity and processing power at several HPCC could be combined on-demand. As a next evolutionary step, the research in Grid computing broadened and became concerned with coordinating the allocation of resource in virtual organizations. This technology is based on virtualization of resources (i.e. processing power, storage capacity, bandwidth, and data). It makes distributed resources available to the user as a single unified system. In general, organizations using Grid technology can optimize the use of their departmental resources by sharing them across departments, run computational-intensive applications on their Enterprise Grid, and even enable collaboration with other organizations [8].

D.J. Veit and J. Altmann (Eds.): GECON 2007, LNCS 4685, pp. 29–43, 2007.

Although the Grid could potentially offer a more efficient way of developing products and creating new business opportunities, the use of the Grid is quite limited at present. Grid Computing is mostly used as a mean for simplifying resource management. Because of that, only a small number of companies are deploying Grid-related technologies and none of the small companies (SMEs) consider using the Grid at the moment [9].

To make the Grid being adopted, companies need to understand the benefits they could gain from using the Grid. The support that they need is a clear analysis of the value chains and the cost models of the Grid. Both will help them understanding the real cost cuts (i.e. the amount of money and time that could be saved using the Grid).

In addition to this, the analysis of incentives and concrete business models is needed. A Grid business model defines a framework for creating new value chains. The analysis of Grid business models will show providers and consumers how to trade resources and software services on the Grid. This opens up opportunities for creating new businesses and revenue streams.

Within this paper, we present the results of our analysis of a set of Grid business models. In particular, we give an overview of existing business models and projects investigating Grid business models in chapter 2. Chapter 3 introduces the taxonomy. It classifies and defines the roles that stakeholders could take on within the Grid. The roles describe atomic functions that could be the basis for new value chains on the Grid. Two examples for using the taxonomy are given in chapter 4. It demonstrates the usefulness of the taxonomy and explains in more detail two abstract business models (i.e. utility computing and software-as-a-service). Finally, the conclusion is given in chapter 5.

2   Classifications of Grid Business Models

2.1   Existing Business Models

The existing Grid business models can be classified according to their origin in research or in commerce. The business models of the research category have mainly been developed by universities and research centers. These business models are based on an open Grid architecture that would allow several providers and consumers to be interconnected and to trade services. The business models of the commerce category have been developed and deployed by a single company with the purpose of selling its own products. These business models usually do not involve several providers.

Research Business Models. The following research projects on Grid business models were examined: GridASP [5], GRASP [6], GRACE [7], and BIG [1]. These projects promote open value chains for trading services on the Grid.

GridASP and GRASP rely on the concept of an application service provider (ASP) for delivering and composing services on the Grid. GridASP offers a scalable and service-oriented architecture (SOA), which offers a lot of potential for creating new products and business models. The business model involves only four roles: the consumer, the service provider (which functions as a portal for the consumer, basically aggregating applications and resources), the application provider (which offers the use of software applications), and the resource provider (which owns hardware resources). From a technical perspective, GridASP addresses all relevant aspects: user management, data and job management, workflow handling, resource brokering, semi-automated application deployment, and security [2][3]. GridASP lacks a better integration of economic functions such as SLA management, negotiation of services, accounting, capacity planning, and pricing.

GRASP also offers a scalable and service-oriented architecture focusing on Web Services and OGSA standards, which offer a good support for service integration. The main focus of GRASP is to allow innovative business models and to integrate economic functions such as accounting, billing, and SLA management into the architecture. As opposed to GridASP, the GRASP architecture offers better support for collaboration between organizations and ASPs by allowing to create virtual organizations and a federation of ASPs.

Compared to GridASP and GRASP, GRACE is very economic-oriented and less focused on architectural issues, aiming to develop a generic framework or infrastructure for a computational Grid economy. The framework provides brokering, service discovery, and trading through an innovative API for negotiating prices and services on the Grid. GRACE relies on existing Grid middleware such as Globus [10] and Legion [11]. GRACE defines only two main roles in exchanging services on the Grid: the consumer, represented by the broker, and the seller or resource owner.

The BIG project addresses the problem of Grid business models from a more general and theoretical perspective [1][4]. It classifies the current Grid projects in four levels based on the support for economical functions and business models. It also focuses on requirements for innovative business models on the Grid and identifies transparency, QoS, brokerage, SLA, and dynamic trust management in virtual organizations to be the most important requirements. BIG supports a large set of innovative applications such as dynamic collaborations, workflows, applications on demand, dynamic resource-management, and resources on demand.

Commercial Business Models. The following commercial models were analyzed: Sun Grid Compute Utility [12], Amazon EC2 [13], the Virtual Private Grid (VPG) [15], and WebEx Connect Application Grid [14]. Both, Sun Utility Grid and Amazon EC2, provide on-demand computing resources while VPG and WebEx provides on-demand applications.

Sun Utility Grid allows the user to create jobs and submit an application, but does not give the user means to control or monitor the execution. The logs and results, together with error reports are provided once the execution is completed. Amazon, on the other hand, allows the user to create virtual machines, which makes the job execution fully transparent. The user has access to virtual machines with 1.7Ghz Xeon CPU, 1.75GB of RAM, 160GB of local disk, and 250Mb/s of network bandwidth. Users can initiate, run, and monitor applications on each virtual machine. Additionally, Amazon provides storage through Amazon S3. For both models, the user is only charged for the consumed resources. Sun Grid Compute Utility charges $1/CPU-hr. Amazon charges $0.10 per instance-hour consumed (or part of an hour consumed), \$0.20 per GB of IP data transferred into and out of Amazon, and \$0.15 per GB-Month of Amazon S3 storage (as of February 2007).

The VPG has been designed by the British Telecom (BT) together with industry partners [15]. The VPG will allow BT to provide its customers with services such as music or video on-demand. The services and resources are developed and are owned by BT's partners. The VPG is based on virtualization of resources and a SOA architecture. This allows combining resources of application and resource providers. However, VPG does not support complex applications such as workflows and has no support for composition of services. Economic functions also lack. Another shortcoming of this model is that BT has full control over the network, and manages the resources and applications, acting as the only reseller. The VPG does not allow other network providers and resource brokers to coexist on the same Grid.

WebEx provides online meeting, web conferencing, video conferencing, and teleconferencing for enterprises. Their software-as-a-service (SaaS) implementation allows innovative and complex, on-demand composition of services and workflows. Moreover, WebEx enables developers to attach new applications to the Grid and sell their products to customers through the WebEx platform. The applications are delivered through their WebEx MediaTone Network, a private global network and platform. MediaTone Network includes connected data centers and servers distributed around the world. However, WebEx only focus on a specific kind of application, namely Web meeting software [14].

2.2 State-of-the-Art in Grid Business Model Classifications

The identification of stakeholders and their roles in Grid business models has been addressed by many scholars [16][17][18][19][20]. Our work harmonizes these classifications and builds our taxonomy of Grid business models (see section 3) on top of it.

Some classifications of classical Internet service providers can be found in [17] and [19]. Their classifications of roles are based on business transactions and are organized in layers. The Software-as-a-Service business was classified in [16]. Another classification of Internet businesses can be found in [18], where the authors illustrate a five-tier classification. Standard bodies, consortia, academic groups of interest, and governments are at the lowest layer, setting the rules for collaboration. This layer is followed by the layer of large technology vendors, niche vendors (who integrate), and application vendors. Within the third layer (influenced by media and information sources), there are consultants, resource service providers, and resellers, which provide customized services to the next layer consisting of business users and retail service providers. This layer provides the provisioning to the last, the fifth layer, consisting of end-users.

In more detail, Grid has been analyzed in [20]. Their model of Grid businesses focuses on the structure of Grid-aware markets. Its layers are divided into two main groups: the Grid market participants and the technology enablers. The Grid market participants consist of three tiers: the service tier (consisting of service providers, content providers, consolidators), the platform tier (consisting of Grid infrastructure providers such as Grid operators and resource providers), and the consumer tier (consisting of partner Grids, virtual organizations (VOs), enterprise Grids, department Grids, and end-users). The technology enablers are also organized into three tiers with the same name as the tiers of the Grid market participants. However, the service tier consists of application providers, the platform tier of middleware vendors, and the consumer tier of consultants and integrators.

3 Roles and Stakeholders

This chapter presents a classification of roles that stakeholders on the Grid could take on. The definition of stakeholders and roles that we follow is: A Grid *stakeholder* is an entity that takes on one or several *roles* in a business model for selling Grid services. A *grid service* is defined as any service that can be provided on the Grid.

Figure 1 shows a classification of roles. There are five categories of roles that a stakeholder could assume on the Grid. Those categories are the roles of a Hardware Resource Service Provider, a Grid Middleware Service Provider, a Software Service Provider, Content Provider, and a Consumer.

I.   **Hardware Resources Service Providers:** This is the lowest layer of the classification representing hardware providers. The hardware can belong to many different providers. In detail, this layer includes:
  A. **Storage Resource Providers:** This role represents the stakeholders providing huge storage systems or a collection of physical or virtual storage resources (located on geographically distributed PCs). Examples are systems such as Amazon's Simple Storage System (S3) or Openomy.
  B. **Computing Resources Providers:** They provide computing resources such as Amazon (Elastic Compute Cloud (EC2)) and Supercomputing centers.
  C. **Network Services Providers:** This layer represents the network providers including ISPs and their multi-tier classification as explained in more details in [17].
  D. **Devices Service Providers:** They provide the access devices such as sensors and microscopes.
II. **Grid Middleware Service Providers:** The stakeholders in this layer provide the services build upon the above-mentioned physical layer. This role is comprised of two distinct roles:
  A. **Basic Grid Middleware Service Providers:** This role represents the collection of stakeholders providing basic Grid functionality. This layer includes the following four main roles:
    a. *Grid Resource Management Service Providers*: This role comprises the stakeholders that provide the following management functionalities:
      1. Resource discovery services.
      2. Resource allocation management services and virtualization.
      3. Resource connectivity services.
      4. Metering and monitoring services.
      5. Job scheduling services.

     b.   *Security Services Providers*: This role comprises the stakeholders involved in providing security functionality.

     c.   *Fault Tolerance Service Providers:* This role consists of error detection, error recovery, job mapping, and check pointing services.

     d.   *Grid Billing Management Service Providers:* This role covers the entire billing stack for any kind of service (1. Accounting services. 2. Charging services. 3. Pricing systems services. 4. *Payment* management services.)

B. Composite Resources Service Providers: This layer represents the role played by stakeholders providing a value-added composite service, which includes services of the previous layers. The stakeholders in this layer take one or more of the following roles:

     a.   *Service Level Agreement Services (SLAs) Providers:* They provide services for the contract (SLA) management, negotiation, monitoring, and auditing.

     b.   *Grid Services Brokers:* Based on the specific task that they provide, brokers can be classified as:

         1.   *Risk Brokers*: These brokers minimize the cost for consumers by finding not only the best deal from several offers based on user specified parameters but also based on the uncertainty of the availability of resources.

         2.   *Trust Brokers*: They help users of the Grid to assess the uncertainty, which results from using resources of unknown providers.

         3.   *Value Brokers:* They perform the task of managing jobs (even entire workflows related to Grid) on behalf of a consumer. This role can also be further sub-divided based on the kind of job (e.g. finding hardware resources and *composing resources).*

     c.   *Capacity Planners:* They assure in the long term that the balance between demand and supply is met. In order to maximize their utility, they calculate how many and when to buy / sell resources.

     d.   *Market-Place Providers:* Stakeholders in this role provide a market place for trading Grid services. These services can be hardware resources, basic grid middleware services, and composite resources.

     e.   *Grid Service Resellers*: They provide selling and retail services of Grid-oriented services.

III. Software Service Providers: A stakeholder in this role takes on one or more of the following sub-roles dealing with software:

A. Application Service Providers: These stakeholders provide (commercial or open-source) application services, either ready to use of-the-shelf packages or customized services. Applications (services) include all types of application such as multimedia, scientific, and business applications. Types of application service providers are:

     a.   *Software-as-a-Service (SaaS) Providers.* An example is the execution and maintenance of an Apache Web server.

     b.   *Software Repository Providers.* This provider maintains a repository of software and controls access to this software.

     c.   *Software Hosting Providers:* This type of stakeholder provides the software *environment* for applications to be executed.

B. Billing Management Service Providers: They are equivalent to the providers of II.A.d.

C. Software Market-Place Providers: Stakeholders in this role provide the market place for software services.

D. Software Brokers: Stakeholders in this role provide the brokering services for software services. They are similar to II.B.b.

E. Software Resellers and Retailers: They sell software services.

F. Applications-to-Grid Wrappers: They provide integration of applications into the Grid.

G. Software Vendors: This stakeholder develops software. They can be:

    a. *Open Source Software Developer:* The source code of this software is available under certain licensing conditions.

    b. *Commercial Software Companies:* This software is mainly proprietary. The use of the software is restricted.

    c. *Software Integrators:* These stakeholders develop software to interface different software components.

IV. Content Providers: This layer represents *roles* available on the information side of our model. Stakeholders in this layer can be divided into the following groups:

A. Content Creators: Any entity that creates content, regardless of the content type (e.g. photo, video, and text).

B. Content Aggregators: They aggregate, classify, and organize content for either business or individual use, using applications such as *mediawiki*, tags as in *clipmarks*, or bookmarking services as *blinklist*.

C. Content Composers: They re-build and modify the content. They add some value and then resell it. An example is the company *programmableweb.com*.

D. Content Distributors: This type includes content update disseminators. Examples are *RSS* services as in *Tailrank* and *Topix.net* as well as traditional content distribution channels such as classical Web media channels.

E. Content Brokers: Stakeholders in this role provide matching services between content providers and consumers.

F. Content Resellers and Retailers: They provide retail services for content (e.g. flicker).

G. Content Market-Place Providers: They provide the market place where content can be exchanged according to some economic rules (e.g. *Flicker*, *YouTupe*, and *Democracy 2.0*).

V. Consumers: This layer represents roles taken on by stakeholders who simply consume a service. Three major types of consumers can be distinguished:

A. Business Users: This entity is either a virtual or physical business entity; seeking value-added services. They can be classified into:

    a. *Core-Business Consumers*: They seek core business process applications. Examples of those stakeholders include: pharmaceutical companies (run applications for drug discovery), financial institutions (run complex applications to make accurate financial estimations).

    b. *General-Business Consumers*: They seek general business process services (i.e. outsourcing of storage, CPU, content development). Examples of stakeholders can be any government agency and business.

    B. End-Users: This entity is a single person or group, which consumes the services without the intention of modifying or reproducing the application.

    C. Universities: This entity has non-monetary objectives when using resources.

In addition to this classification, we need to mention the emergence of the following three roles. These roles provide additional services for stakeholders, spanning over the above-described layers (Figure 1):

- *Grid Consultants:* This role provides consultation service (such as economic analysis, technical analysis, education, and training) for Grid adopters. Examples of stakeholders are software integrators and strategic consultants.
- *Grid Standardization Bodies*: This role provides standardization services, which could be taken on by stakeholders such as academia, governments, and consortia.
- *Regulators*: This role will guide the development of the Grid through policies.

To complement our classification of roles on the Grid, we need to mention the following four facts about stakeholders that take on those roles:

- Any provider can offer integrated services through *horizontal service integration* (i.e. integration roles of the same layer) and / or *vertical service integration* (i.e. integration of roles of different layers). The integration of Grid technology, Web services, and Web2.0 enables this. It will give each stakeholder the potential to change his role by adding / deleting more roles to his stake and build new business models. It will allow all stakeholders to adapt quickly to new market conditions.
- Service providers can become consumers of services and vice versa.
- Not all of the providers mentioned have to be present in the future market. Each of the providers can serve a different niche market.
- Even though the layered structure is the ideal case, the stakeholder relationships between the layers can follow different paths. The path that a consumer takes to use a service does not have to go through all the layers. The consumer can directly choose the most appropriate service from any layer. However, the services of different layers have to be ordered according to the layered structure.

Figure 1 summarizes the roles of the stakeholders and depicts the relationships between them. The arrows represent the direction of service delivery. They are used to indicate the service that a stakeholder delivers to another stakeholder in a different role. The following chapter analyzes two business models using this taxonomy.

4 Role Analysis of Two Reference Business Models

We discuss two abstract business models with respect to the service functionality that they require for their implementation. The first reference business model is "Economically Efficient Utility Computing". In this case, the user owns the software that will be executed on the Grid. The second reference business model is "Software-as-a-Service", which explains the value chain of software on demand. In this case, the user rents the software and has the option to specify the hardware resources on which the software should run.

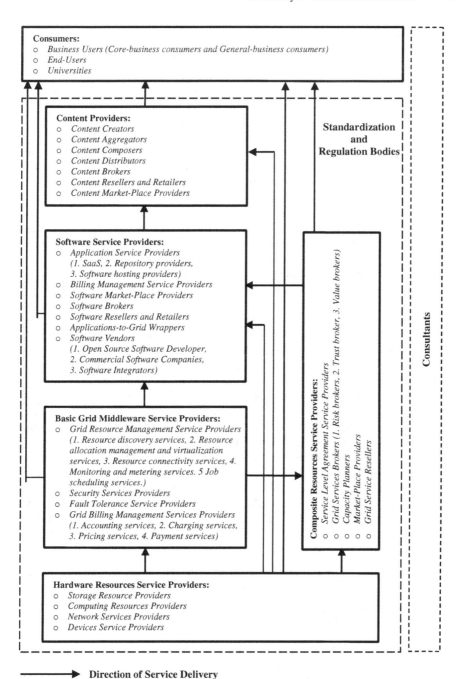

Fig.1. Roles of stakeholders and their relationships

4.1 Reference Business Models: Economically Efficient Utility Computing

There are a number of advantages of utility computing. First, different Grid systems that provide utility computing can perform load balancing amongst each other, thus ensuring that the capacity of the Grid is used to the highest extent possible. Second, utility computing implies that computational power is always present. The Grid is inherently fault-tolerant. (It is unlikely that all single Grid systems will fail on the Grid at the same time.) Furthermore, the operation of the Grid is transparent, meaning that the user is not aware of the fact that the application is running on a geographically distributed system. All these advantages do not only enable the execution of computationally intensive, scientific applications but also allow commercial customers to use the power of such a Grid to solve their problems quickly and efficiently.

However, there are many different kinds of users (e.g. SMEs, large enterprises, the general public, and academia), distinguishing themselves in the amount of budget, urgency of their application, and quality of service expectations. For example, industry users, which try to achieve a competitive advantage, require the termination of the execution of their application within a specific period of time. Because of that they are willing to pay a higher price than other users. In order to get their job executed on time, i.e. get priority over other users, the Grid could charge them a premium fee, which is specified in an agreement between the user and the Grid system. This agreement, which is called a Service Level Agreement (SLA), states what the user wants and what the provider promises to supply in a legally binding way. An SLA also describes a measurable performance standard and a penalty fee if it is not delivered.

This paradigm of utility computing supports the execution of any application (i.e. high-performance computing workflows, parallel applications, or simple sequential application, such as Web servers). Any end-user can get access to any kind of computational resource, ranging from supercomputers, which provide large computing power, to a single PC. Figure 2 shows the interaction between the consumer and the Grid. In this case, the consumer owns a software application and executes it on the Grid. The Grid offers the mechanisms for deploying and executing the application (e.g. automatic deployment, execution monitoring, and hardware resource discovery). Figure 2 also illustrates the business processes (or service interactions) for purchasing hardware resources on the Grid and for executing the application. Implementing such a business model requires at least the following basic roles, which belong to three layers:

- **Hardware Resource Service Providers:** For the utility computing business model, server, storage, and network resources are considered.
- **Grid Resource Service Providers:** This intermediate layer between the Consumer and Hardware Resource Service Providers offers only services needed to execute an application:
  - **Basic Grid Middleware Service Providers:** The basic Grid middleware must provide at least security, and accounting.
    - **Billing Stack Service Provider:** Interacts with the Resource Broker to charge the Consumer for consumed resources. It has a Pricing component for storing current and past prices and components for accounting, charging, and billing. When the execution completes, it bills the consumer.

□ **Grid Resource Management Service Providers:** It provides virtualization, metering, job deployment, and resource discovery.

o **Composite Resource Service Providers:** In this example of utility computing business models, only one type of provider is needed:

□ **Resource Broker:** It selects the best-fitting resources from different Hardware Resource Service Providers upon user request and initiates the job deployment and monitors the job execution.

• **Consumers:** The consumer runs its application on the Grid, using the services of the Grid Resource Service Provider. The consumer can be of any type.

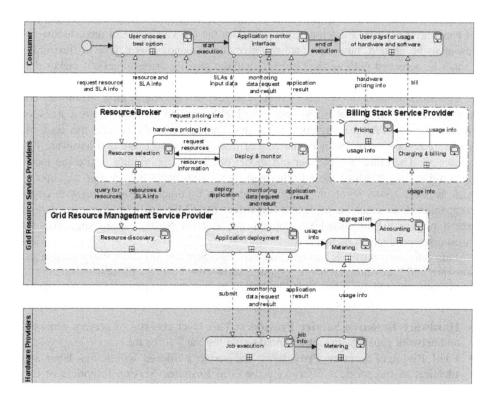

Fig. 2. Interaction of roles in the utility computing business model

A typical scenario of this business model can be found in high-performance computing. Applications in this scenario require huge amount of hardware resources. Applications range from scientific (simulations such as weather and climate modeling, weather prediction), digital media (animation, special effects, rendering), life sciences and health care (drug discovery, structure-based design, molecular dynamics, medical imaging), financial services (Monte Carlo simulations, risk analysis) to manufacturing.

4.2   Reference Business Model: Software-as-a-Service (SaaS)

For software owners, this model has a number of advantages. First, they no longer need to be concerned with license agreement violations, which are common when software is sold directly to the customer. Second, since consumers pay only for their usage, even consumers, who cannot afford buying very expensive software, can now purchase units of software usage. These additional customers will increase the amount of income for the software owner and, thus, contribute to higher profits from software development. However, the following third reason is the most important one for software vendors. Since the software vendor can use the Grid to run its software, they do not need to own the hardware resources themselves. Instead, they can reserve a set of hardware resources for a long time period on the Grid and, should more hardware resources be needed, buy additional resources on demand. This degree of flexibility cannot be achieved without the Grid (i.e. without sharing of resources). It makes the Grid an attractive alternative to buying and maintaining hardware resources.

The SaaS business model also has advantages for customers. First, customers do not have to buy additional, sometimes even highly specialized hardware resources to run purchased software anymore. This does not necessarily reduce the cost of hardware resources, but reduces the cost of ownership of software and hardware for customers. Therefore, the pay-per-use model opens the opportunity to use the most powerful software, which would otherwise be too expensive to buy. For customers, such as SMEs, SaaS levels the playing field when it comes to competing with large companies. This aspect is especially important for SMEs in fields like metallurgy, which require highly complex computations, and for SMEs, which specialize in customized products in niche markets.

This reference business model, Software-as-a-Service, involves the purchase of software and hardware resources. Figure 3 shows the business process of running a SaaS business model on the Grid. The basic services that a system has to provide to offer software-as-a-service belong to four layers:

- **Hardware Resource Service Providers:** This layer consists of servers, storage, and network capacity that are required to run the SaaS software.
- **Grid Resource Service Providers:** The services offered within this layer are identical to the services offered by the Grid Resource Service Providers of the utility computing business model.
- **Software Service Providers:** For the SaaS business model, this layer represents service providers who maintain software, which they do not necessarily own and execute on Grid resource hardware. The providers considered here are:
    - o **Software Discovery Provider:** Software vendors have registered their applications in a Software Registry. Information about available software applications and their SLA can be retrieved from the registry.
    - o **Software Broker:** It uses a Software Discovery Service to retrieve information about similar applications that match the consumer's preferences. It then offers a selection to the consumer.
    - o **Application Service Provider:** In the SaaS business model case, it is the environment, in which the SaaS software is executed on the Grid.

   ○ **Billing Management Service Provider:** Interacts with the Application Service Provider and with the Billing Stack Service Provider of the Grid Resource Service Provider layer to provide services for billing the consumer for the consumed hardware resources and the application usage. When the execution completes, it bills the consumer.

- **Consumers:** This entity is the consumer, who uses the software. It buys access to the software on a usage-basis, using the services provided by the Software Service Providers. It can be a SME or an individual.

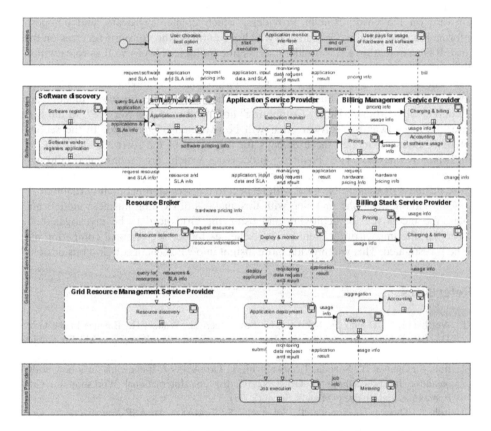

Fig. 3. Interaction of roles in the software-as-a-service business model

The SaaS business model allows SMEs to gain access to expensive commercial software that they could not afford to purchase licenses for or because they do not have the expertise to develop the application inside the company. Another interesting scenario is the one of starting a business with little investment by using the software services (or a composition of them) provided on the Grid and adding own expertise. For example, an interior decorator could use rendering and visualization software to provide advice on house decoration.

5   Conclusions and Future Work

This paper presented a survey of current Grid business models. It then identified the roles, which can be assumed by different stakeholders on the Grid, and classified those roles into five groups, defining the taxonomy of Grid business models. The groups are: Hardware Resource Service Providers, which own storage, network capacity, devices, and server capacity; Grid Resource Service Providers, which provide the Grid middleware and composite resource services; Software Service Providers, providing the software and the environment for managing and executing software on the Grid; Content Providers, which create, aggregate, and compose information; and, finally, Consumers, which are entities who consume services on the Grid. It is to note, that a stakeholder can assume multiple roles as well as roles within different groups.

In addition to this, we discussed two reference business models and applied the taxonomy to those two business models. These reference business models are "Economically Efficient Utility Computing" and "Software-as-a-Service". In the first case, the user owns the software that will be executed on the Grid. In the second reference business model, the user pays for the usage of the software and the Grid hardware resources. The business processes, which are represented through interactions between roles, give a general guideline for what is needed to implement those business models on the Grid with respect to the service functionality required.

After having made the first step with this analysis of existing business models and roles, the incentives for deploying and using the Grid need to be investigated. In particular, the impact of different pricing schemes for service-oriented computing has to be investigated. This will provide more insight into value chains of Grid businesses.

References

1. Weishäupl, T., Donno, F., Schikuta, E., Stockinger, H., Wanek, H.: Busines In the Grid: The BIG Project.GECON2005. In: The 2nd International Workshop on Grid Economics and Business Models, Seoul (2005)
2. Itoh, S., Ogawa, H., Sonoda, T., Sekiguchi, S.: GridASP - A framework for a new utility business. In: proceedings of GECON2005, the 2nd International Workshop on Grid Economics and Business Models, Seoul (2005)
3. Itoh, S., Ogawa, H., Sonoda, T., Sekiguchi, S.: GridASP: An ASP Framework for Grid Utility Computing (2005),
   http://www.teragrid.org/programs/sci_gateways/docs/gridasp-ogawa.pdf
4. BIG Faculty of Computer Science, University of Viena (2005), Web site
   http://www.cs.univie.ac.at/project.php?pid=79
5. GridASP (2007), Web site http://www.gridasp.org/en/
6. GRASP (2007), Web site http://eu-grasp.net
7. GRACE (2007), Web site http://www.gridbus.org/ecogrid/
8. IBM Grid Computing (2006), Web site
   http://www-1.ibm.com/grid/about_grid/what_is.shtml
9. Schikuta, E., Donno, F., Stockinger, H., Vinek, E., Wanek, H., Weishäupl, T., Witzany, C.: Busines In the Grid: Project Results (2005),
   http://www.pri.univie.ac.at/Publications/2005/Schikuta_austriangrid_bigresults.pdf

10. Globus (2007), Web site http://www.globus.org/
11. Legion (2007), Web site http://legion.virginia.edu/index.html
12. Sun Grid (2007), Web site http://www.sun.com/service/sungrid/
13. Amazon Web Services (2007), http://www.amazon.com/b/ref=sc_fe_c_1_3435361_1/104-3351913-5483951?ie=UTF8&node=201590011&no=3435361&me=A36L942TSJ2AJA
14. WebEx Connect, First SaaS Platform to Deliver Mashup Business Applications for Knowledge Workers (2007), http://www.webex.com/pr/pr428.html
15. Falcon, F.: GRID – A Telco perspective: The BT Grid Strategy, GECON2005. In: The 2nd International Workshop on Grid Economics and Business Models, Seoul (2005)
16. Traudt, E., Konary, A.: Software-as-a-Service Taxonomy and Research Guide (2005), http://www.idc.com/getdoc.jsp?containerid=33453&pagetype=printfriendly#33453-S-0001
17. Altmann, J.: A Reference Model of Internet Service Provider Businesses. ICTEC2000. In: 3rd International Conference on Telecommunication and Electronic Commerce, Dallas, Texas (November 2000)
18. Bryson, M.: What To Do When Stakeholders Matter: A Guide to Stakeholder Identification and Analysis Techniques. London School of Economics and Political Science (February 2003)
19. Altmann, J., Routzounis, S.: Economic Modeling for Grid Services. e-Challenges 2006, Barcelona (October 2006)
20. Plaszak, P., Wellner Jr., R.: Grid Computing: The Savvy Manager's Guide. Elsevier, Amsterdam (2006)

Development of a Generic Value Chain for the Grid Industry

Katarina Stanoevska-Slabeva[1], Carlo Figà Talamanca[2], George A. Thanos[3], and Csilla Zsigri[4]

[1] Mcm Institute of the University of St. Gallen, Blumenbergplatz 9, 9000 St. Gallen, Switzerland
Katarina.Stanoevska@unisg.ch
[2] Innova S.p.a.; via G. Peroni 386, 00131 Roma, Italy
c.talamanca@innova-eu.net
[3] Network Economics and Services Group, Athens University of Business and Economics, 76 Patission Str. Athens,Greece
gthanos@aueb.gr
[4] Atos Research&Innovation, c/ Llacuna 161. planta 3, 08018 Barcelona, Spain
csilla.zsigri@atosresearch.eu

Abstract. Grid middleware provides the fundamental framework for the provision of Grid services. However, Grid middleware is complex software that consists of several modules. The modules for a specific grid middleware exhibit in many cases complementary features and are produced by different software providers. Thus, in order to provide a complete grid solution for business customers, it is necessary to establish a complete value network comprising all relevant suppliers. Business aspects of grid, such as business models and value networks have not been considered broadly in research yet. This paper contributes to fill this gap by describing the value network of a grid case study and by aggregating the results into a generic grid value chain.

Keywords: Grid business models, Grid value networks, Business Grids.

1 Introduction

Under grid we understand a specific middleware, which provides the necessary functionality required to enable both sharing of heterogeneous resources and virtual organizations [1]. Up till present, research in grid has mainly concentrated on technical aspects and development. In addition, the initial and core application area of grid technology is eScience. There are ongoing international (for example EGEE) and national initiatives that are dedicated to developing and running grids in specific data and processing intensive scientific areas.

The business market of grid (i.e. the market of grid services for companies) has not been fully exploited yet. Based on first successful examples from the eScience application area, grid technology is entering a new level of maturity and is getting productized with the aim to enter the corporate market [2]. In order to enter the corporate market, suitable business models are required [2], [3]. Providers of grid

D.J. Veit and J. Altmann (Eds.): GECON 2007, LNCS 4685, pp. 44–57, 2007.

solutions need to evaluate suitable value chains, pricing and licensing approaches, and market development and entrance strategies [2]. Up till present, less attention has been paid to these aspects of grid technology. Economic oriented research questions have only been considered in the research area "Grid Economics" [4]. Grid economics considers, however, the application of economic paradigms for resource allocation on the technical level (for example application of auctions for market distribution of available resources in a grid) and does not consider the research questions regarding creating successful business models for a profitable market entrance of grid technology. Thus, there are no guidelines available for grid technology providers, such as how to choose the appropriate business model and value chain [2], [3].

This paper provides a contribution to the business-oriented research of grid. It focuses on value networks of grid solutions for the corporate market. The findings regarding major players on the grid market and their relationships resulting from a broad literature review and in-depth case studies of the value chain of 18 representative grid industry pilots are aggregated to a generic value network for the provisioning of grid services and products.

The content of the paper is structured as follows: section 2 provides an overview of the research approach. Section 3 comprises a description of the grid case study. Section 4 contains a description of the concept for a generic grid value network. Section 5 concludes the paper with a summary and outlook.

2 Research Approach and Definitions

2.1 Value Chains and Value Webs - Definition

A first step in the construction of a business model is the study of the process of creating and exchanging value. The analysis of the value creation system helps organizations to understand how the different entities work together to produce value. The value chain analysis is a very efficient tool for tracing product flows, showing the value adding stages, identifying the key actors and the relationships with other actors in the chain.

The value chain analysis goes back to Porter's traditional linear model of value chains [5]. The liner model of value chain that consists of a sequence of value-enhancing activities has been an important and sufficient instrument for analyzing the value creation process in a company or industry during the last century. However, in the current networked economy relationships among companies are more complex and value creation is rather multidirectional than linear [7]. Given this, the linearity of the value chain proposed by Porter impedes the correct understanding of key processes such as relationships, alliances, and partnerships among the involved firms. Among the most important assets that are exchanged in the network are not only the monetary flows but also knowledge, trust relationships, intellectual property and leadership [6]. Several concepts have been presented that extend the concept of value chain towards value networks in the literature. For example Tapscott et al. [6] propose the concept of value web. Pil and Holweg [7] propose the concept of value grid. Further terms to denote the concept of extended value chain are: value network, business web and similar. In this paper we will use the term value network. Based on an aggregation of elements of different definitions for value networks [6], [7], the term value network will be defined in this paper as follows: A value network is a web of relationships that generates economic value and other benefits through complex dynamic exchanges between two or more individuals, groups or organizations.

A value network analysis can be performed from the perspective of a company or an industry. An industry-level value network serves as a model of value creation and relationships in the industry. It is composed of all the value creating activities within the industry. To identify the aggregated value network of an industry requires a good understanding of the complementary products and services to provide a complete solution and what kind of relationships among players are present or possible.

2.2 Research Approach

In order to identify the major players and their relationship in the grid market the following research approach was followed:

- In a first step a broad literature survey on the subject was conducted.
- In a second step an in depth-analysis of the value web of 18 representative grid industry pilots was performed. For the purposes of the paper one example belonging in the financial sector is presented. All the grid pilots form part of the Integrated Project (IP) "BEinGrid" (http://www.beingrid.com/) that is funded by the European commission under FP6. One of the main objectives of the European project BEinGRID is to consider and develop in systematic manner a repository of knowledge and guidelines regarding business and market aspects of grid technology. In the heart of the project there are 18 business experiments that are piloting grid technology in various key industrial sectors such as the textile, gaming, ship-building, film-making, logistics, and retail management industries.
- In a third step the findings from step 1 and step 2 have been aggregated to produce a generic value network. The generic value network can potentially be applied by grid technology providers to position them and to evaluate which partners they need.

2.3 Results from the State-of-the-Art Research

The state-of-the-art analysis revealed that the topic of business models and value networks for grid technology has not been considered broadly yet. On the one hand there are several market studies available that are provided mainly by market research institutions [8], [9], [10], [11] or papers that elaborate on potential diffusion and adoption strategies for grid in enterprises [14]. These types of publications provide either descriptions of concrete cases or a general overview of the market and diffusion potential for grid technology. On the other hand there are first examples of papers considering the market entrance of grid technology, but on a general level [13], [14]. All papers identify in general two major business models for grid technology:

- Selling grid technology as combined software and consulting product.
- Providing grid enabled application and grid services according to the paradigm Software as a Service (SaaS).

The first business model is structured around grid software as a specific software product that can be offered either in a commercial manner or in an open source manner. However, the transition to grid computing in companies is a major endeavor that requires considerable changes in existing processes, application and governance of the information infrastructure [2], [15]. Thus, an indivisible part of grid software is respective consulting for companies how to master that transition.

The second business model is based on the SaaS paradigm [16]. SaaS is a relatively recent model of software access. It builds on the latest advances in technology within the software industry in order to offer a radically different model for accessing and using software. As the name states, SaaS is a way of accessing software products as services. This is significantly different to the traditional means of accessing software and raises a number of problems, both from the technical and legal perspectives. In this model a user can combine services or even software components (as in the service-oriented architecture paradigm) from different Grid providers and build his service. Providers on the other hand can provide their software in different packages and prices to meet the customer needs. Software can be accessed remotely and run over the grid infrastructure of the provider. This is in contrast to the traditional software model where software would be purchased from a retailer, generally in a box with a manual and some storage media containing the software binaries. SaaS makes software accessible according to a service/utility model.

The research question related to potential and available value networks for specific grid solutions for different application areas was considered in several research projects:

- The Akogrimo project (http://www.mobilegrids.org/) proposes a consolidated value chain for grid in mobile application [16].
- The project GRIDEcon (http://www.gridecon.eu/) explores potential value networks for different grid scenarios [17].
- The project GridASP (http://www.gridasp.org/wiki/) focuses on value networks for grid utility computing, i.e. for the SaaS business model [18].

However, the value chains and networks proposed within the projects are either dedicated to a specific application area or specialized for a specific business models. The most comprehensive study about grid value chains for both type of business models described above was conducted by Forge and Blackmann [2]. They propose the following value chain (c.f. 1):

The above value chain provides a structured overview of the specific competences, products and services necessary to provide a complete grid solution for the business market. It is a valuable picture of how grid services are assembled to a complete product. However, it does not provide an overview of involved players and how the competences and value adding activities are divided among them. It also does not provide information about the relationships among involved players and it is not granular enough to provide a basis to understand how competences can be bundled. Given this, the value chain proposed by [2] and the other projects mentioned above has provided the basis for an enhanced value network including actors as well as their competences and relationships presented at the end of this paper.

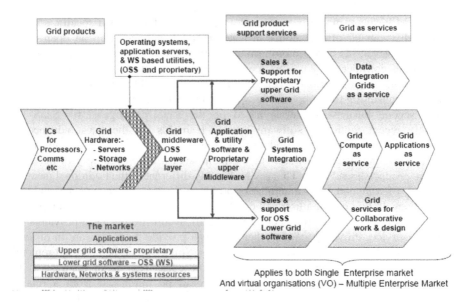

Fig. 1. The grid value chain according to [2]

3 Example of a Value Network in a Grid Pilot for Financial Portfolio Management

3.1 The Motivation of the Financial Industry for Grid in Portfolio Management

Over the past decades portfolio and risk management techniques have adapted to increasingly complex financial instruments and risk scenarios. The rapid growth in derivative financial instruments and the derivatives losses reported in recent years have intensified concerns regarding reliably measuring financial instrument risk exposure.

In this respect, one of the biggest challenges is the surge in data volumes that have to be manipulated for risk and performance calculations. Market conditions as well as compliance requirements - both of which require additional sources of information to be included in the risk calculation - are the major reasons for this increase in volumes. Consequently, the risk computation cycle time increased significantly, almost stretching into the start of next day's trading cycle. As a result, the operational risk in this environment also increased considerably.

The financial market is under considerable and mounting pressure for more transparent and reliable risk reporting. To meet this demand, managers need a whole risk and evaluation infrastructure at their fingertips and this implies systems, technology and data. It is clearly an issue that is essential to all financial products - such as bonds, options, credit products and structures- to access the full credit curve and the well defined volatility surfaces. From the managers´ point of view, the key is to generate meaningful risk reports; nevertheless, this requires new technological solutions and high computational resources.

3.2 The Involved Players

The aim of the grid pilot considered in this paper is to develop an application to run simulations on a grid infrastructure to support financial institutions in strategic decisions of financial portfolio management problems. According to a pre-defined level of risk, the tool will calculate the portfolio with the best performances. The involved players in the value network are:

User: This business pilot has two end users. They present two different ways of operating in the financial bank sector, which highlight different needs and applications for the new tool. One of them will implement the new application to manage the customers' portfolios, while the other will use the new tool to optimize the asset allocation of proprietary capital. In a commercial situation, the two users will use the new grid application to support decision-makers in the asset allocation of portfolio management problems.

Service provider: The service provider in the considered case is at the same time the technology provider and offers the following:

- Computation power through super computers and computational infrastructure.
- Grid software services built on top of the grid middleware (Grid portal, Data and Storage Services, Information Service).
- Consultancy, expertise and competencies on placing legacy applications on a grid environment and to design grid-aware applications as well as specific expertise on the design of the software architecture for the portfolio management application.

Application provider: The competencies of the application provider involved in the pilot are related to the design and implementation of quantitative decision support tools for several application fields. The methodologies adopted belong to different areas of mathematical programming, such as machine learning, stochastic programming and simulation/optimization.

This market player offers its own experience in the design of effective and efficient quantitative tools to support end-users in the Portfolio Management applications. In particular, it assists end-users in the definition of mathematical representation of specific applications and provides the kernel to deal with the different phases of the decisional process, from a mathematical standpoint. The developed solutions exploit the advantages provided by the grid technology.

In a commercial situation, the role of this party consists in consulting and providing additional services on demand, like customization and tuning of the decision support system for additional requirements. Moreover, the provision of enhancements of the mathematical model and of the solution approach, according to the academic results in this field.

System integrator: The integrator's role is to act as interface between the end-users and the tool developer. In the pilot, this party participates in defining the requirements of the new application and assists the end-users in the data gathering activity and during the pilot phase. Therefore, it fosters the whole process of transforming the early pilot into a new business, promoting it successfully in the marketplace.

In a real business scenario, the integrator has two main roles: (i) promotion and commercialization of the new application and service through its network (reseller); (ii) consultant for business process reengineering to integrate grid applications and personnel training.

The value and contribution of the Integrator is the support to the end-user for the integration of grid technologies in its business processes, and also the re-design of the business models. The value of a new grid application depends from how well it is integrated in the business processes and how effectively it is used. A re-engineering of the business processes and even the re-design of the business models can increase the value of the same grid application. In fact, the economic viability of a new application depends not only on its intrinsic value, but also on its efficient use and proper integration.

3.3 The Value Network of the Pilot

During the pilot phase, the value network is reduced to the basic supplier–end-user relationship, with the service provider running the grid infrastructure and the new application, and the financial institutes representing the end-user and simulating the different portfolio combination. Figure 2 describes the same value network in a commercial environment. The relationships among the involved players are more

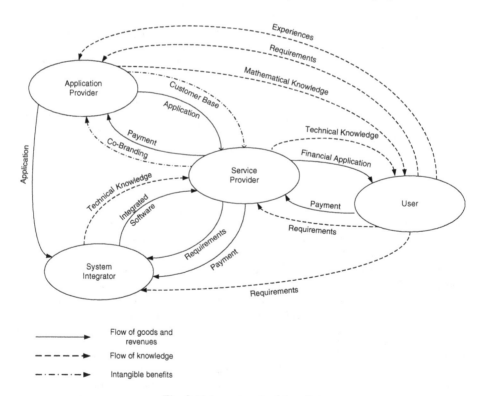

Fig. 2. Value network of the pilot

complex and multi-faced. The service provider is the "main interface" to the customer and bundles the services of the application provider and the system integrator. He hides the complexity of the solution for the end customer and resolves potential problems as synchronization of licensing strategies and bundling of different component into a complete solution. Considerable intangible advantages can be achieved for each involved player in the value network. For example the application and service provider can leverage on each others customer base. In addition, the system integrator provides technical know-how to the solution provider and can leverage his customer network.

3.4 Future Potential Value Chains

Another possibility would be to extended the above described value network, by inclusion of additional players and by a different distribution of value creation activities among involved players. A multiplicity of new actors can enter in this simple value map, differentiating the services and creating new business models. For example, as described in figure 3, an intermediate financial service provider (FSP) can enter the network. The FSP owns an extensive database of financial data uses the grid infrastructure as a resource and sells on demand customized forecasts on financial portfolio arrangements to the banks and financial institutes. In this scenario the "main interface" to the customer is the FSP and the complexity of the technical solution is hidden from the end customer.

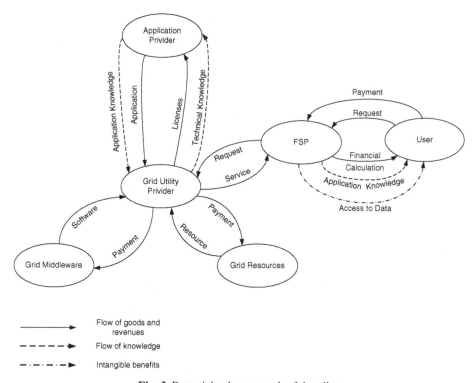

Fig. 3. Potential value network of the pilot

4 Integration of Results in a Generic Value Chain

In the previous section a value network of a specific grid pilot was presented and analyzed as an example. In a similar manner the remaining 17 Business Experiments of the BEinGrid project were analyzed. The findings were aggregated to the BEinGrid consolidated value network. The BEinGRID consolidated value network has been produced based on the identification of all the actors that appear in the different business experiments of the project and contribute to the creation of value and the interactions among them. The basic idea is to show how the content is distributed across a net of market actors reaching the different industries. Based on the generic value chain, specific value networks can be created.

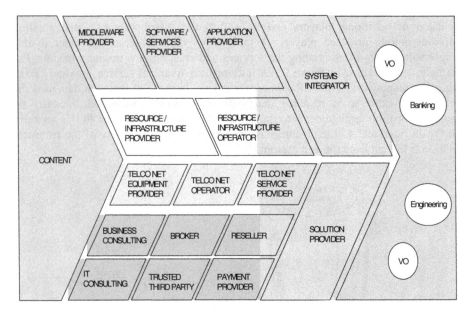

Fig. 4. The BEinGRID consolidated value chain

The following table describes the roles of the actors of the BEinGRID consolidated value chain:

Table 1. BEinGRID consolidated value chain roles description

Role	Description
Content (provision, aggregation, distribution)	Data, information and experiences created by individuals, institutions and technology to benefit audiences in contexts that they value. Content to be processed and transformed to build the final "product". The end-user can provide the content.

Table 1. (*continued*)

Grid middleware provider	Provides libraries, executable codes that implements the grid functionality. (standards + software (lower software + upper software)).
Software/services provider	Provides software that is usually added to platforms or targeted to special niche markets. (e.g.: Independent Software Vendor (ISV)) An ISV makes and sells software products that run on one or more computer hardware or operating system platforms. The "Service Provider" offers services that run on the technology in question. These Service Providers will likely have a strong relationship with the Application Providers or with the operators. The main idea behind this business participant is that external service provider can offer their services to operators and application providers.
Application provider	Is the first customer of a specific platform. The Application Provider can buy a development package to integrate its software on top of the respective technology. An application service provider (ASP) is a business that provides computer-based services to customers over a network. Software offered using an ASP model is also sometimes called on-demand software.
Resource/Infrastructure provider	Provides equipment (hardware) on which the grid implementations run. Other hardware, network and system resources used (e.g. HP).
Resource/Infrastructure operator	Provides access to and use of the equipment that it owned by the resource or infrastructure provider.
Telco network (equipment) provider	Provide equipment (telco hardware and network resources) that build the telco network (e.g.: Nokia, Siemens).
Telco network service provider	Sells bandwidth under specific business criteria. Many times the network service provider and the network operator is the same company (Telefonica, BT, Vodafone).

Table 1. (*continued*)

Telco network operator	Provides a broadband communication network, offering real time functionality and easy access. Enabler of communications. It also can play the role of end user when the grid technologies are used for the company business processes.
Business consulting	Offers a solution to your business problem, optimizes your processes, improves your "numbers" telling you how, and provides business models, advices you in business development and marketing. (e.g.: Accenture, Atos, Logica).
IT consulting	Expertise for assistance in the IT (information technology) processes, computing services, training.
Payment provider	Provides infrastructure and management enabling the payment transactions between actors. It can be a financial entity, a business consulting company, a broker, a network service provider, etc.
Reseller	Companies that resell/distribute an existing solution provided by another company. It can be the whole suite, or one or more of its components.
Broker	Intermediary, can also be the trusted third party. It advices on you on which grid solution fits better to your situation. Provides services based on specific quality levels required by the end-users.
Trusted third party	Deals with contractual arrangements, financial settlements, and authentication of users (e.g.: a bank or other financial entity).
Systems integrator	Integration of the different modules (software, hardware) required to build the grid solution. Brings the players together. Technical role, but may also do consultancy work besides installation, deployment and IT support.
Solution provider	Offers you a package of network, middleware and applications (e.g.: IBM). It may provide also consulting or grid expertise so that the solution of the problem can be determined.

Table 1. (*continued*)

Market	Targeted end-users or virtual organizations (VO) in different industries.

The list of potential players provided above shows that the value network for grid solutions is quite complex. Given this it might be organized in sub-clusters or networks that are represented by a lead player that bundle the offerings of several players and join them with the offerings of other players to a complete solution. The potential clusters that can form on the market are (c.f. 5):

One possible cluster can be lead by the systems integrator, who integrates the services and offerings of the application, middleware and resource provider to a grid application. Thereby he bundles the offerings, and resolves potential conflicts regarding licenses and pricing.

Another cluster - the telecom cluster - can be formed by the providers of services and equipment necessary to enable communication infrastructure for the solution. This type of cluster can be lead by the network operator or network service provider.

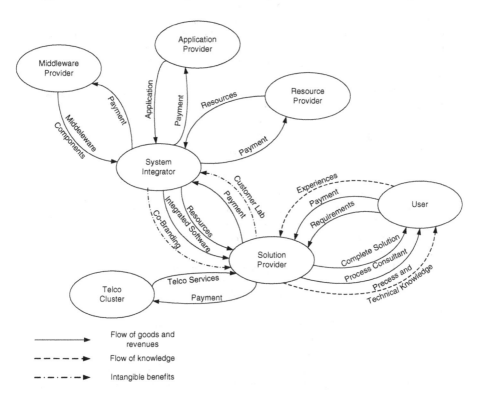

Fig. 5. The BEinGRID generic value network

Finally the offerings of the system integrator and the teleco cluster can further be enriched with consulting services by the solution provider, who is the main interface to the customer.

5 Summary and Conclusion

The aim of the paper is the analysis of existing value chains and development of a generic value network for the grid industry. In order to achieve this, an in depth state-of-the-art analysis was performed. Then, based on a case study existing and potential value networks were developed. Finally, the results from step one and two were combined in a generic value chain for the grid industry. The generic value chain provides an overview of actors and their competences. It can be applied by providers of grid solutions or components for grid solutions to position themselves and to identify potential necessary partners to enter the business market.

References

1. Foster, I., Kesselman, C., Tuecke, S.: The Anatomy of the Grid - Enabling Scalable Virtual Organizations. In: International Journal of Supercomputer Application (2001)
2. Forge, S., Blackmann, C.: Commercial Exploitation of Grid Technologies and Services - Drivers and barriers, Business Models and Impacts of Using Free and Open Source Licensing Schemas. Final Report of the European Study Contract No. 30-CE-065970/00-56
3. Timmers, P.: Business Models for Electronic Markets. International Journal on Electronic Markets and Business Media 8(2), 3–8 (1998)
4. Buyya, R., Abramson, D., Venugopal, S.: The Grid Economy. Found, under (2005), http://citeseer.ist.psu.edu/buyya05grid.html
5. Porter, M.E.: Competitive Advantage: Creating and Sustaining Superior Performance (1985)
6. Tapscott, D., Ticoll, D., Lowy, A.: Digital Capital - Harnessing the Power of Business Webs. Harvard Business School Press, Boston (2000)
7. Pil, F.K., Holweg, M.: Evolving From Value Chain to Value Grid. In: MIT Sloan Management Review, vol. 47(4), pp. 72–80 (Summer 2006)
8. The 451 Group: Does Sun have a future as a grid service provider (March 10th, 2005)
9. Quocirca: Business Grid computing - the Evolution of the Infrastructure. Homepage of Quocirca. Business & IT Analysis. Found (September 13th, 2006), under http://www.quocirca.com/pages/analysis/reports/view/store250/item1515/
10. The 451 Group: Grid Computing - Where is the value? 451 Grid Adoption Research Service, Report 1 (August 2004)
11. The Insight Research Corporation: Grid Computing - A VerticalMarket Perspective 2005-2010. Executive Summary. Found (2005), under http://www.insight-corp.com/reports/grid06.asp
12. Schikuta, E., Donno, F., Stockinger, H., Vinek, E., Wanek, H., Weishäupl, Th., Witzany, Ch.: Busines In the Grid: Project Results. In: Proceedings of the 1st Austrian grid Symposium (2005)

13. Sawhny, R., Dietrich, A.J., Bauer, M.Th.: Towards Business Models for Mobile Grid Infrastructures - An Approach for Individualized Goods. In: Proceedings of Practical Aspects of Knowledge Management, Vienna (2004)
14. Joseph, J., Ernest, M., Fellenstein, C.: Evolution of Grid Computing Architecture and Grid Adoption Models. IBM System Journal 43(4), 624–644 (2004)
15. Geiger, A.: Service Grids - von der Vision zur Realität. In: Barth, T., Schüll, A. (eds.) Grid Computing. Konzepte, Technologien, Anwendungen. Wiesbaden: Friedr. Vieweg & Sohn, pp. 17–32 (2006)
16. Hafner, M.: The Akogrimo Consolidated Value Chain. Business Modelling Framework Dissemination Level, WP 3.2 (2005)
17. Stockinger, H.: Grid Computing: A Critical Discussion on Business Applicability. IEEE Distribute Systems Online, vol. 7(6), Art. No. 0606-o6002 (2006)
18. Ogawa, H., Itoh, S., Sonoda, T., Satoshi, S.: Concurrency and Computation: Practice & Experience. GridASP: an ASP framework for Grid utility computing 19(6), 885–891 (2007)

Strategies for the Service Market Place

Paul McKee[1], Steve Taylor[2], Mike Surridge[2], Richard Lowe[2], and Carmelo Ragusa[3]

[1] BT, Adastral Park Martlesham Heath, Ipswich IP5 3RE, UK
paul.mckee@bt.com
[2] University of Southampton IT Innovation Centre, 2 Venture Road, Chilworth, Southampton
SO16 7NP, UK
{sjt,ms,rl}@it-innovation.soton.ac.uk
[3] Multimedia & Distributed Systems Lab (MDSLab), Department of Mathematics,
University of Messina, Contrada di Dio 98166 S. Agata – Messina, Italy
cragusa@unime.it

Abstract. We describe a number of strategies for a future service oriented market place. We describe the SLA's role within the service framework, and how it enables customers to make value judgements regarding the quality of a service. We also discuss the complexity of too much choice from both the customer and provider points of view, and advocate a "discrete offer" approach. We discuss the "cost of negotiation" and argue that it must be carefully balanced with the cost, value and risk of the offering being negotiated for. We add to the negotiation analysis with presentation and discussion of some results showing a simulated Grid market place and show that it is possible for service providers to deny themselves work through attempting to offer a high quality guaranteed service.

Keywords: Service Oriented Architecture, Service Oriented Infrastructure, Grid Computing, Business, Service Level Agreement.

1 Introduction

Businesses are faced with increasing pressure to introduce products and services in increasingly shorter time scales and at reducing costs. This has led to great interest in technologies such as service oriented architectures (SOA) and its supporting service oriented infrastructure (SOI), grid computing, and the new range of software as a service (SAAS) offerings. Using such technologies it is hoped that businesses will be able to rapidly create new applications, either for use internally or for sale, from services that may be assembled and executed rapidly and economically. To achieve maximum economic benefit it is essential that the services are reused and that, as the technology matures, fewer are developed from scratch. This all leads to the supposition that there will be a market place for applications and services that will have a global reach and potential consumers of services will have to choose from such a global marketplace. In this paper we present some observations about the structure and characteristics of such a service marketplace and suggest strategies that we believe will lead to greater user satisfaction.

D.J. Veit and J. Altmann (Eds.): GECON 2007, LNCS 4685, pp. 58–70, 2007.

For the purposes of this paper, we define a service as: "the doing of work for the benefit of others". We use this generic definition so as to concentrate on the business aspects of the provision of service rather than a technical definition (for example a Web Service). By definition services are intangible, and may not even exist before they are required. This means that the potential purchaser has limited opportunity to assess the quality of a service before use. Hence there is a need for the consumer to determine the nature of a service and whether it will live up to their expectations. This may be achieved through the use of service level agreements (SLAs), described next.

2   The Role of SLAs

Within the 6th Framework IST project NextGRID we have proposed the use of bipartite (two-party) SLAs to describe both the functional and non-functional aspects of a service to allow consumers to make informed choices. An SLA is an agreement (as denoted by its name) and a contract for service is likely to be based on it, so the SLA describes to what each of its signatories is obliged to do in order to be compliant with the agreement (see [8]). The SLAs under development in NextGRID (also see [13]) are significantly different from what could be regarded as conventional SLAs in two main ways:

- NextGRID SLAs are expressed in terms of *customer business benefit*; and
- NextGRID SLAs may impose *interdependent obligations* on either party.

These differences are discussed in detail below.

We believe it is essential that the SLA have a strong end user benefit component, and that services being offered should be described in terms of their impact at the business level of the consumer, and not in terms of the technology supplied. The marketing benefits of this should be obvious – it is much easier to convince a financial decision maker of the benefits of a service that guarantees one of his key business processes using agreed metrics rather than one that guarantees the availability of, for example, a number of servers. We believe that the strong linkage to business impact benefits both the customer and provider, and this strong business linkage in the SLA structure is discussed in a previous publication [16]. The customer can then assign and manage costs transparently within their business and the service provider retains the flexibility in operations to strive to reduce the cost of delivery and ultimately the price paid by the customer. This view contrasts with the use of SLAs in the more established academic Grid community where the SLAs described in [2], [6] and [20] focus on resources (e.g. computers, network bandwidth, storage devices) rather than services. This is of course appropriate for a community of experienced users running often experimental codes. However, in order to address the needs of business (as opposed to technical) users the value of the service must be articulated at the appropriate – business – level rather than the resource level. The customer should not be concerned with the resources required to provide the service, just that the service exists and provides clear business benefit – management of resources should remain the domain of the service provider. Buco et al [1] concur with this view: *"...The customer need not know the implementation details of the provider's service level management (SLM) processes..."*. Buyya et al [2] propose a "Grid Economy" in which the focus is on the resources, low-level computational elements such as CPU, network and storage. Our position is that the value to the customer is at a higher level than the hardware. For example, application codes contain much more customer value than hardware, as they embody functionality that can solve customers' problems.

The second major characteristic of a NextGRID style SLA is that it recognises the fact that in a future service oriented marketplace application delivery will be collaborative and that SLAs need to recognise the impact of this collaboration. We advocate the notion of *interdependent obligations* that describe necessary pre-conditions in order for the SLA to be valid, and that some obligations will be dependent on others. For example, if a customer wishes to outsource payroll processing and asks a service provider to guarantee that the payroll processing will always be completed by the 29th of any month, it is entirely reasonable that the service provider place a pre-condition on the customer requiring that data be available 24hrs before the delivery deadline. The WSLA specification [19] also denotes the use of obligations, but does not determine that they may be dependent relationships, and how these relationships are dependent. Through the use of obligations the NextGRID SLA clearly sets out the expectations and obligations of both parties involved, and we believe is an essential component of a future electronic market place for services.

3   The Dynamic Service Market Place and Choice

Those who market products and services often assume that the more choice given to a customer, the more likely they are to find something that satisfies their requirements. At first glance it seems that this assumption is valid in a future service-based market place. We believe however that trying to offer a wide range of choices may turn out to be less successful than expected, and even counterproductive.

Consider the role of a service provider in such a market. In order to deliver a service against an SLA the service provider will need to characterise the performance of the service in various operational conditions and at all combinations of the various quality parameters the customers may choose. For this, the service provider must produce automated management tools to ensure performance is maintained for all permutations. This is of course extremely difficult, expensive and slow. The customer will see increased costs to pay for these developments, and a reduction in the number of new and innovative services due to the complexity of the task. Although limiting the number of available products may seem counterintuitive, the advantages for controlling costs may be significant. Gottfredson and Aspinall [10] discuss the notion of the "innovation fulcrum" and its impact on a number of manufacturing activities. They describe how careful selection of options and minimisation of complexity leads to a more efficient business and that each business has an optimal number of products. If a business offers too few products they may miss out on customers, while offering too many products leads to greater complexity and unacceptably high management costs. A balance between adequate market coverage and limitation of complexity must be sought.

As a service consumer the problems associated with too much choice may be equally severe. Faced with a service that contains a large number of potentially variable terms the end user is unlikely to be aware of the trade offs associated with combinations of the various settings, and they may unwittingly specify a set of parameters that significantly increases the cost of the service without increasing the quality. Indeed the availability of a large number of parameters may in fact obscure which parameters are in fact important to them – as discussed by Twing [18] it is essential that SLA focus on critical service items that drive the business, and have clear financial implications. We regard this as further evidence supporting the need for SLAs to be described in terms of business benefits to the consumer. As the end user is a person, other non-technical problems may be observed. Iyengar and Lepper [14] have shown that when a customer has too much choice they do not always see this as a benefit and often they become less likely to buy anything at all, and the wider range of options often leads to dissatisfaction with the final choice. Customers often resort to very simple and non-optimal strategies to choose in such conditions, and this raises an interesting question for automated systems in that if the customer is unable to make the choice effectively in person, how can they program an automated system to achieve the best result for them?

4 Discrete Offers – The Supermarket Approach

If we consider the 20th century approach to shopping much of the complexity has been removed from commodity purchases by some degree of standardisation. Many household purchases have standard pack sizes that facilitate both automated production (thereby reducing costs) and efficient comparison between products by the consumer. It is fairly easy for the customer to make a value judgement between two packets of different branded soap powder, for example. In addition, the prices of the goods would greatly increase if the customer had total freedom to specify their features. For these reasons, within the NextGRID project we have advocated the use of a "discrete offer" system for services. It should be noted that by "discrete offer" we do not mean one single offer, but that every service provider should offer services with fixed quantised service levels, each constituting a separate, "discrete" offering of their service. Customers may then have the confidence that the combination(s) of quality parameters offered will at least work, and that the service provider has created a suitable management infrastructure for each level of the service, hopefully leading to reduced costs.

We recognise the fact that the commoditised approach will not be suitable for every situation and that there will be a continuum stretching from the simple purchasing of commodity services through to more complex fully negotiated contracts. The deciding factor governing which strategy is appropriate will be the cost incurred in the negotiation process versus the cost of the delivered service and this is discussed in the next section.

5 The Cost of Negotiation

The cost of negotiation comprises a significant part of what have been termed "transaction costs", the roots of which were proposed in 1937 by Coase [5]. Transaction costs are (financial and otherwise) costs associated with the transaction itself (for example costs associated with the process of procurement), not the item or service that is the subject of the transaction. We believe that the cost of negotiation must be carefully managed. Negotiation cost must be insignificant compared to the cost of the item or service being negotiated for. It is only worth spending much time, money and effort negotiating when the cost, value and risk associated with the item or service being negotiated for are significant.

In line with the discussion in Section 0, the supermarket approach has no negotiation. The customer chooses from a set of discrete offers and pays the labelled price. This is appropriate because the vast majority of real-world supermarket items are low-cost and commoditised. Conversely, a high-value, high-cost, complex service contract will have a great deal of negotiation, as the end item is extremely expensive and the risks of not carefully working out the contact between the customer and provider are significant. In the real world, a contract for the construction of a football stadium would not go ahead without careful negotiation from all parties concerned. The costs, value and risks are too great to leave negotiation to chance.

Applied to the Grid, there are a number of automated "discover, find and bind" protocols and demonstration systems (for example those described in [7] and [17]) some of which use software agents to perform the negotiation to gain access to the services being negotiated for. Whilst these are certainly useful, we assert that there must be a limit determined by the cost, risk and value of the negotiated item, above which humans must get involved. An organisation may be perfectly happy for software systems to take responsibility for low-value agreements, but humans must get involved when the stakes get higher, so that they can make more advanced decisions and take responsibility for their actions on behalf of their organisation. The greater the level of risk and investment, the greater the need there is for human decision making and responsibility. You cannot sue a computer because it made a promise your organisation cannot deliver!

There are a number of factors that determine the total cost of negotiation, and measures that can be taken to control costs arising from each.

1. There is a significant time cost to the customer of searching for suppliers. This implies that customers should search for suppliers infrequently, and record any approved suppliers in an "approved supplier list".
2. Many negotiation protocols use an "invite-tender" approach. This is where an invitation to tender is issued by a customer and providers respond to it. There is usually a significant delay in the issuance of the invitation and the delivery of the tenders. The delays make this a high-cost approach, and in general this should only used for high-cost bespoke services. It is useful for building long-term supply relationships e.g. when deciding who to include in an approved supplier list (see above).
3. The customer will have difficulty in comparing the value from offers from different providers – these are not likely to be directly comparable and thus the customer has to compare apples and oranges to determine which suits their needs best. This cost can be controlled by standardising the terms used in discrete offers – an approach used successfully by the supermarkets (see Section 4).

4. As discussed in Section 0, there is a significant risk to customer and provider of giving the customer too much freedom in choosing the service profile – what they choose may not be workable or be too costly to provide. Providers should restrict customers to an appropriate number of discrete offers.
5. There is a considerable time and risk cost to both customer and provider of iterative negotiation. Offers and counter-offers need to be computed and responded to, and there is the significant risk that there is no guarantee that iterative negotiation will converge. Providers can avoid this cost by making only discrete offers, allowing customers to choose between them but not to negotiate over their terms.
6. There is a considerable financial cost of any human interaction (decision making, responsibility, etc). Humans are much more costly than computers, and cannot handle large numbers of negotiations. Service providers and ideally also consumers should automate their procedures for managing offers and making agreements.

In conclusion, providers and customers need to analyse the costs and potential risks of any negotiation they enter into and appropriately target resources. Depending on the situation, too much or too little attention to negotiation and customer choice can be costly or dangerous.

6 Self Denial of Service

In this section we briefly describe simulation results of an initial study into the behaviour of a "Grid market place". The overall aim of this work was to arrive at a means of simulating or modelling a Grid market place, given that there was no real-world information regarding how different actors operated in such a situation.

This work builds on earlier work in the GEMSS project [9]. This investigated use of the WSLA schema [19] for SLAs along with FIPA negotiation protocols [15], in a business to business Grid middleware based on GRIA [11] (the current GRIA software downloadable from [11] uses the discrete offer approach and does not employ FIPA protocols). The GEMSS middleware was deployed and evaluated using only a small number of sites. Here, we aimed to find out how such a B2B approach would work on a larger scale, with full market competition between larger numbers of sites.

Before starting the work, three options were considered for understanding Grid marketplace behaviour: direct measurement, analytical modelling and simulation. We chose a simulation based approach, as it was deemed to be the most flexible. Measurement was obviously not possible, and it was deemed that simulation allowed more "what-if" types of analysis to be performed than modelling.

We based the simulator code on the GridSim toolkit [3], [12] with appropriate extensions. There are two main actors in the simulation – users and service providers. The business goal for users is to get some (compute-based) work performed within a deadline. Hence, for the users we added a negotiation capability with service providers. The users issue requests for work consisting of a *workload* and a *deadline* by which the workload needs to be processed, evaluate responses from providers, and select the most appropriate based on a utility function. The workload is specified in units relative to a hypothetical "standard machine", and all actors in the simulation use the same metric and units to represent the demands of users and the capacity of providers. When evaluating offers from providers, the users differentiate between the offers using a *cost-time* selection strategy. The cost-time algorithm is described in [4], and has price as the major factor but run-time is taken into account where providers cost about the same.

The business goal of the service provider is to provide compute services that process the users' workloads within the required timeframes. For the service provider we added two major extensions. Firstly, we added the capability to negotiate with users. The providers receive requests and make offers, containing the commitment to perform work for the user and a price. The offers expire after a specific time (denoted as O_p – the *offer validity period*), and users may only accept them within this time. Secondly, we eliminated the disclosure of status by service providers. In the original GridSim, the users are aware of the internal resource utilisation of the service providers. In a market situation, this is clearly not desirable for the service provider, as it gives the user an advantage in any negotiation. Here, the only information users are aware of is whether a provider can complete the work inside a given deadline, and what it will cost the user. There is no communication of the resources used by the provider, only whether the provider can satisfy the user's requirements.

The scenario chosen models the following situation. In a future large scale dynamic market place it will be trivial for potential customers to request quotations from large numbers of suppliers at frequent intervals in the hope of spotting price trends, or windows of opportunity for low cost operation. It is easy to imagine automated software continually evaluating the providers for bargains. If the service providers commit their resources to offers when making those offers, so as to provide a high-quality, "guaranteed" service, they will be exposed to the significant risk of self-denial of service through committing resources for offers that are not taken up.

We chose a population of 80 users and 20 service providers for this scenario. The ratio between the number of clients and service providers is based on the assumption that in a marketplace of this type (SME clients and larger service providers) there will be significantly more clients than service providers. The population used here was dictated by the need to have enough providers to allow some variation between them, but not so many users that the simulation was impracticably slow.

The service provider population comprises different types, categorized in terms of the number of processing units, the price, and the speed of the processing units:

- **Size:** Big (9 units), Medium (3 units) and Small (1 unit);
- **Speed:** Fast (20 times a "standard" e.g. client machine) and Slow (10 times a standard machine); and
- **Price:** varies between 2½ cents and 7½ cents per hour on a 'standard' machine.

To create service providers' profiles, all combinations of the above categories were initially considered. We then eliminated the non-realistic ones – for example we can see that a small, low priced service provider is not realistic because it will not create enough revenue to survive. We also excluded slow, low capacity, expensive service providers, since they are very unlikely to succeed in a competitive market. The service providers' distribution thus reflects a marketplace in which large service providers exploit the economies of scale to reduce their costs and therefore attract the majority of the business. However, we have included some medium and small providers as well to study their impact on the market. The final service providers' profiles are shown in Table 1.

Table 1. Service providers' profiles & parameters

Size - Speed - Price	Number of service providers	Number of processors	Processor rate (multiple of standard processor unit)	Price per processor - hour	Equivalent standard processor - hour price
Big - Slow - Low	3	9	10	€0.25	€0.025
Big - Fast - Low	2	9	20	€0.50	€0.025
Medium - Slow - Low	3	3	10	€0.25	€0.025
Medium - Fast - Low	1	3	20	€0.50	€0.025
Big - Slow - Very High	1	9	10	€0.75	€0.075
Big - Fast - High	2	9	20	€1.00	€0.050
Medium - Fast - High	3	3	20	€1.00	€0.050
Small - Fast - High	5	1	20	€1.00	€0.050

We used one key principal for service providers' operation. When making an offer, the service providers commit resources sufficient to process the workload requested by the user and by the user's deadline. The offer is valid for the duration of O_p. After this time, any resources committed for offers that are not accepted are released and can be offered again. Until then, resources committed to an offer may not be used to back up a new offer, thus ensuring that the service providers do not over-commit their resources.

The simulations were used to investigate the effect of varying O_p on the distribution of work across different providers. Figure 1 shows the situation where O_p is 3456 seconds and shows the amount of outsourced work processed by each of our population of 20 service providers (described in Table 1). In Figure 1, the service providers are denoted by a number and code:

SIZE{Big|Medium|Small} SPEED{Slow|Fast} PRICE{Low|High|V.High})

Due to the users' *cost-time* minimisation strategy (minimise cost then use time to sort any equal-cost providers) we expect the cheapest service providers to get most work. The fastest should also achieve the highest utilisation, since their resources are committed for the least time, and they should be most often in a position to make new offers. Figure 1 shows this effect: SP01 to SP05 inclusive are highly utilized, and within this set the fast providers (SP04 and SP05) are more highly utilised.

Naively, we would expect the work to be shared amongst all the low-priced providers, but we are seeing here that SP09 gets very little and SP06-SP08 get none, even though they are all low-priced. Yet some expensive providers (SP11 and SP12) are winning a great deal of work (they are above 80% utilised).

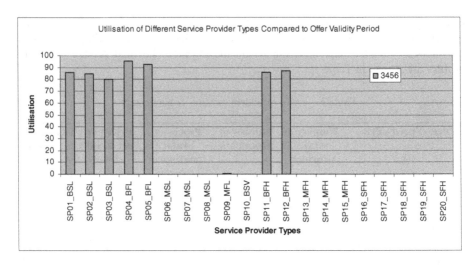

Fig. 1. Utilisation for different service provider types for O_p = 3456 seconds

This is due to the relative agility of these service providers. When providers make offers, they must commit resources for as long as the offers are valid. Hence the low-capacity (medium size and below) service providers commit all their resources much sooner than the larger ones and as a consequence cannot make as many offers as higher capacity providers. Thus the lower capacity providers are considerably less agile and unable to compete, even though their low price should make them attractive to customers.

If we reduce O_p we get a significant change in the work distribution. Figure 2 shows the results of a simulation where O_p was reduced to 1440 seconds.

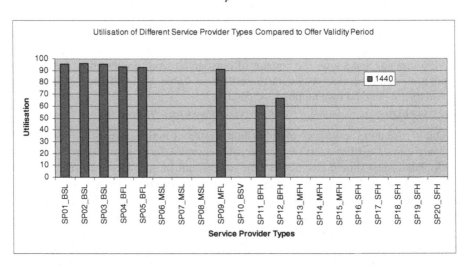

Fig. 2. Utilisation for different service provider types for O_p = 1440 seconds

The major change is that now SP09 (medium size, fast and low-price) is winning much more work (it is now over 90% utilised). SP09 was previously prevented from winning much work because its resources were fully committed, and by the time they became free the more expensive but more agile SP11-SP12 had won most of the work. SP09 is now only committing its resources for 1440 seconds, and is thus able to make more offers and win much more work. The expensive providers SP11 and SP12 still do quite well, but lose out whenever SP09 is in a position to compete with them.

In Figure 3, O_p is further reduced, this time to 720 seconds, and we see another shift in the work distribution following the trend of the lower capacity workers winning work.

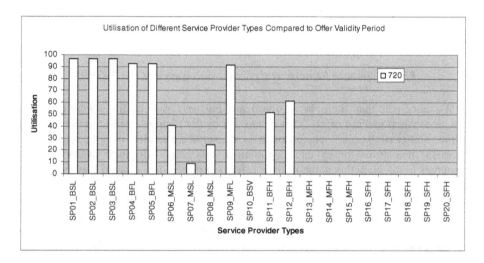

Fig. 3. Utilisation for different service provider types for $O_p = 720$ seconds

SP06-08 are now starting to win against SP11 and SP12. Even though they have very low capacity, the further reduction of O_p means that they can make enough offers in order to stand a good chance of some getting accepted.

Note that low capacity providers become more viable with shorter offer validity periods, even though the more expensive but more agile competitors also use the shorter offer validity period. We did not simulate different O_p for each provider, but it is clear that there will be some advantage in cutting it relative to competitors, though only to the point where customers find it too short.

Note also that shortening O_p further does not further improve the situation for the low-agility providers. By optimising the simulation it became possible to reduce O_p to only 5 seconds. Figure 4 shows that although SP06 to SP08 do a little better, they still cannot achieve the same utilisation as the more agile and expensive SP11 and SP12.

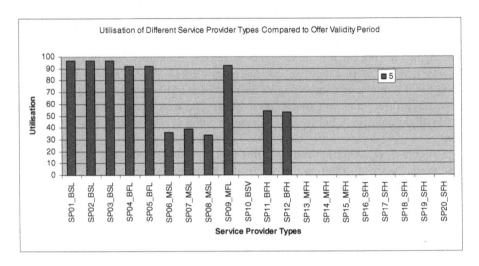

Fig. 4. Utilisation for different service provider types for Op = 5 seconds

We conclude that there is a significant business risk to the commitment of resources when making offers, as it makes a provider highly sensitive to the duration of the validity of the offer. The problem for the service providers is in the relationship between their provisioning and negotiation strategies – do they overbook their resources in the hope that some of their offers get turned down, or do they price their service accounting for the offers that are not taken up?

This is common in resource-intensive real business. It is well-known that airlines routinely overbook their planes to account for "no-shows", and the terms of the contract they have with their customers states that the customers paying the cheaper prices (usually in economy class) may not get a seat on the plane (this is known as "bumping").

It seems clear that the best way to operate services is by targeting a range of customer groups with a range of different service offers that provide different levels of guarantee (sometimes none at all) at different prices. These would have to be encoded in machine readable SLAs so negotiation can be automated on both sides. The SLAs will need to be much more sophisticated than hitherto, as there will be times when services are not taken up, or are taken up but then not delivered. This also implies a more sophisticated approach to manage the services and resources, in which failure to deliver service is an option, and decisions will be needed as to which services should be maintained. Clearly, there is a need to reduce complexity as much as possible, which suggests that the discrete offer approach should be used by service providers. It is also clear that commitment of resources during negotiation will add to the complexity, and that negotiation protocols should not depend on this.

7 Conclusions

In this paper, we have considered strategies for service oriented future markets. This work has come out of discussion, research and thinking into what is currently called the "Grid", and looks like it will evolve into a service oriented market, and hence the emphasis on services in this paper.

We have discussed the difficulty consumers have in judging the quality of an intangible an un-experienced service. The well-known solution is the service level agreement (SLA), and we describe the SLA's role within the service framework. Our chief findings are twofold. Firstly, an SLA and service providers' offerings must use terms understandable to and clearly having benefit for customers. Secondly, an SLA should contain interdependent obligations that *all* its signatories (not just the service provider) agree to.

We have advocated a "discrete offer" approach for service providers. We argue that if customers are given too much choice, they are likely to get overwhelmed by the permutations available to them, and may produce unworkable combinations of service factors. In addition, it has been argued that there is an "innovation fulcrum" - an optimum point between adequate market coverage and enough simplicity and manageability.

We have discussed the cost of negotiation and argue that it must be carefully balanced with the cost, value and risk of the offering being negotiated for. We have taken this further using simulations of a Grid market place, which show that it is possible for service providers to deny themselves work through attempting to offer a high quality guaranteed service. This illustrates that 'transaction costs' are not limited to the actual cost of a transaction process, but may also need to take account of the impact on business agility and opportunity costs. We conclude by determining that there should be different levels of service related to price targeting different customer needs, and presented to customers as discrete offers, with minimal negotiation and minimal pre-commitment of resources until an SLA is agreed.

Acknowledgments. The authors acknowledge the funding of the European Commission, and the work reported here was conducted within the NextGRID (The Next Generation Grid) Integrated Project, Contract Number 511563. This paper expresses the opinions of the authors and not necessarily those of the European Commission. The European Commission is not liable for any use that may be made of the information contained in this paper.

References

1. Buco, M.J., Chang, R.N., Luan, L.Z., Ward, C., Wolf, J.L., Yu, P.S.: Utility computing SLA management based upon business objectives. IBM Systems Journal 43(1), 159–178 (2004)
2. Buyya, R., Abramson, D., Venugopal, S.: The grid economy. In: Proceedings of the IEEE, vol. 93(3), pp. 698–714 (2005)
3. Buyya, R., Murshed, M.: GridSim: a toolkit for the modeling and simulation of distributed resource management and scheduling for Grid computing. Concurrency and Computation: Practice and Experience 14(13-15), 1175–1220 (2002)
4. Buyya, R., Murshed, M., Abramson, D.: A Deadline and Budget Constrained Cost-Time Optimization Algorithm for Scheduling Task Farming Applications on Global Grids. In: Proceedings of the 2002 International Conference on Parallel and Distributed Processing Techniques and Applications (PDPTA'02), June 24 - 27, 2002, Las Vegas, USA (2002)

5. Coase, R.: The Nature of the Firm. Economica 4(16), 386–405 (1937)
6. Czajkowski, K., Foster, I., Kesselman, C.: Agreement-Based Resource Management. Proceedings of the IEEE 93(3), 631–643 (2005)
7. Czajkowski, K., Foster, I., Kesselman, C., Sander, C., Tuecke, S.: SNAP: A Protocol for Negotiating Service Level Agreements and Coordinating Resource Management in Distributed Systems. In: Feitelson, D.G., Rudolph, L., Schwiegelshohn, U. (eds.) JSSPP 2002. LNCS, vol. 2537, pp. 153–183. Springer, Heidelberg (2002)
8. Davis, Jr., T.H., Fitzgerald, P.K.: Deconstructing Service Level Agreements. New York Law Journal (March 4, 2002)
9. GEMSS project, http://www.gemss.de/
10. Gottfredson, M., Aspinall, K.: Innovation Versus Complexity: What IS Too Much of a Good Thing? Harvard Business Review 83(11), 62–71 (2005)
11. GRIA: www.gria.org
12. GridSim toolkit, http://www.buyya.com/gridsim/
13. Hasselmeyer, P., Qu, C., Schubert, L., Koller, B., Wieder, P.: Towards Autonomous Brokered SLA Negotiation. In: Cunningham, P., Cunningham, M. (eds.) Exploiting the Knowledge Economy: Issues, Applications, Case Studies, IOS Press, Amsterdam (2006)
14. Iyengar, S.S., Lepper, M.R.: When choice is demotivating: Can one desire too much of a good thing? Journal of Personality and Social Psychology 79(6), 995–1006 (2000), http://www.columbia.edu/~ss957/whenchoiceabstract.htm
15. Middleton, S.E., Surridge, M., Benkner, S., Engelbrecht, G.: Quality of Service Negotiation for Commercial Medical Grid Services. In: Journal of Grid Computing, Springer, Heidelberg (to appear, 2007)
16. Mitchell, B., Mckee, P.: SLAs A Key Commercial Tool. In: Cunningham, P., Cunningham, M. (eds.) Innovation and the Knowledge Economy: Issues, Applications, Case Studies, IOS Press, Amsterdam (2006)
17. Shen, W., Li, Y., Ghenniwa, H., Wang, C.: Adaptive Negotiation for Agent-Based Grid Computing. In: Falcone, R., Barber, S., Korba, L., Singh, M.P. (eds.) AAMAS 2002. LNCS (LNAI), vol. 2631, pp. 32–36. Springer, Heidelberg (2003)
18. Twing, D.: Are you savvy about SLA negotiations? Network World (May 10, 2005)
19. WSLA: http://www.research.ibm.com/wsla/
20. Yeo, C.S., Buyya, R.: Service Level Agreement based Allocation of Cluster Resources: Handling Penalty to Enhance Utility. In: Proceedings of the 7th IEEE International Conference on Cluster Computing (Cluster 2005), IEEE Computer Society Press, Los Alamitos (2005)

Prediction-Based Enforcement of Performance Contracts

Thomas Sandholm[1] and Kevin Lai[2]

[1] KTH – Royal Institute of Technology
Center for Parallel Computers
SE-100 44 Stockholm, Sweden
sandholm@pdc.kth.se
[2] Hewlett-Packard Laboratories
Information Dynamics Laboratory
Palo Alto, CA 94304, USA
kevin.lai@hp.com

Abstract. Grid computing platforms require automated and distributed resource allocation with controllable quality-of-service (QoS). Market-based allocation provides these features using the complementary abstractions of proportional shares and reservations. This paper analyzes a hybrid resource allocation system using both proportional shares and reservations. We also examine the use of price prediction to provide statistical QoS guarantees and to set admission control prices.

Keywords: Admission Control, Proportional Share, Computational Market.

1 Introduction

Grid applications traditionally run on dedicated machines, with a fixed performance level that depends on the hardware configuration. In this model, the main source of uncertainty in predicting job deadlines is the queue waiting time. As a solution to heterogeneity, and low resource utilization various virtualized platforms are emerging, such as Xen, VMWare, and VServer. In a virtualized Grid, where the performance level is configured dynamically based on job requirements and current demand, the main source of uncertainty is the risk of not being allocated enough capacity. The allocation decisions are complicated by the scale, and distribution of the Grid resources, and the vast variability and complexity of the job requirements. Therefore, it is not feasible to make these decisions manually using static configurations or policies.

Market-based allocation is one form of allocation that is automated, distributed, and provides QoS. Market-based allocation supports two primary resource abstractions: proportional shares and reservations. A pure proportional share allocator always admits new resource requests and continuously reallocates resource shares in response to the current load. This fully utilizes the resources and always admits well-funded resource requests, but may cause an earlier request to fail a minimum resource requirement. In contrast, a pure reservation allocator fixes resource shares at purchase time. Admitted resource requests in a reservation system will always (modulo failure) meet their

D.J. Veit and J. Altmann (Eds.): GECON 2007, LNCS 4685, pp. 71–82, 2007.

resource requirements, but sometimes utilization is low, and sometimes well-funded requests will be rejected admittance.

In this paper, we examine a hybrid system that mixes both proportional share and reservation abstractions to achieve the best of both worlds: satisfying quality-of-service requirements for some applications while maximizing utilization and providing resource availability for latecomers. Using simulation, we explore how such a hybrid system performs for different workloads.

In addition, we examine how prediction algorithms affect the result. Prediction of future load is critical to efficient resource allocation. Proportional share allocators require it so that purchasers can get statistical QoS guarantees. Reservation allocators require it to set the prices for reservations. However, the effect of universal prediction on a system is not obvious. For example, if low prices are predicted for a particular hour of CPU time, then many resource consumers may try to buy it, thus ruining the accuracy of the prediction.

We base this analysis on previous work on predicting demand in computational markets [1,2], where we evaluate different prediction techniques to give accurate percentile bounds for expected demand for arbitrary probability distributions. We assume here that we have an approximation for the cumulative distribution function (CDF) of the demand. Furthermore, we assume a computational market where proportional share resource allocations are enforced (e.g., Tycoon [3]).

Our contribution in this work is twofold: 1) we highlight and visualize issues with statistical guarantees in performance contracts using simulations, and 2) we propose and implement a solution to these issues using contract admission control.

The paper is organized as follows: Section 2 provides an overview of the mathematical models used to analyze and simulate our resource allocation scenario, Section 3 presents and discusses the design and results of our simulations, Section 4 reviews related work, and finally Section 5 sums up our findings with some concluding remarks.

2 Model

2.1 Statistical Guarantees

We are interested in analyzing what bids individually rational resource consumers should place on their tasks, given that they need a certain performance level to finish within a deadline. Different guarantee-levels can then be compared based on the price consumers have to pay for obtaining a performance level.

To formalize the model we use the following standard probability theory notations:

$$x \in X, P(x) = P(X = x) \tag{1}$$

$$D(x) = \int_{x_{min}}^{x} P(\varepsilon)d\varepsilon \tag{2}$$

where P is the probability function (a.k.a. PDF), and D the probability distribution function (CDF). To find performance levels based on guarantees it is also useful to look at the inverse of the distribution function, or percent point function (PPF), defined as:

$$D^{-1}(D(x)) = x \tag{3}$$

The proportional share resource allocation model is defined as:

$$q = \frac{b}{b+c} \tag{4}$$

where q is the performance level or QoS in terms of resource share $(0,1)$, given a consumer's bid, b, and a measured price, c, of a resource. The price is the sum of all existing bids on the resource.

A rational consumer would hence bid

$$b = \frac{cq}{1-q} \tag{5}$$

for any measured price, c, to maintain a service level q. However, in a competitive computational market the price adjusts dynamically to the resource demand, and can thus be viewed as a random variable C, which changes continuously over time. Since, q depends on c it can also be seen as a random variable, Q. The guarantee of delivering a certain QoS level to the consumer, g, will be expressed in terms of this random variable Q.

$$q \in Q, c \in C \tag{6}$$

$$g = P(\frac{b}{b+C} > q) = P(C < \frac{b}{q} - b) = D_c(\frac{b}{q} - b) \tag{7}$$

where D_c is the price distribution function. Now using the inverse of the price distribution function we can calculate the bids to place given a QoS level and a guarantee

$$D_c^{-1}(g) = b(\frac{1}{q} - 1) \tag{8}$$

which gives

$$b = \frac{D_c^{-1}(g)}{\frac{1}{q} - 1} = \frac{D_c^{-1}(g)q}{1-q} \tag{9}$$

The intuition behind this is that the probability of getting a service level greater than a certain value is the same as the probability of the price being below a particular value, or

$$P(Q > q) = P(C < c) \tag{10}$$

2.2   Admission Control

Now, we would like to offer an admission control service with more than a statistical guarantee for an additional fee. We calculate this new price as:

$$b' = \frac{D_c^{-1}(g+r)q}{1-q} \tag{11}$$

where b' is the price a user needs to pay to get share q with guarantee g, and r is the fee parameter. Note that the fee is not simply added to or multiplied with the bid, but included in the percent point calculation of the price. This ensures that the admission control service is more expensive when there is a high price difference in offering a higher guarantee, in order to account for the expected loss the provider makes when refusing new consumers due to admission control.

In our model, a share of a resource can be requested with either an absolute guarantee paying the admission control fee, or with a statistical guarantee paying the spot (current) market price. The admission controller makes sure that no request is accepted that violates previously admitted requests with absolute guarantees. Whether a violation would occur as a result of admitting a new request is determined by enumerating and evaluating bids and required shares for all active previously admitted requests for the same resource. Consequently, all requests for the resource will need to go through the same admission control path in order to ensure reservation-like guarantees. We note that price volatility in this model is paid for directly by the user, and the admission controller operates in the interest of the provider to keep the prices at a higher level to compensate for not being able to preempt existing low-paying allocations in the event of higher-paying requests. Alternatively, the admission controller could be separated entirely from the resource being provisioned and operate like an insurance agent to put in spot market bids on the resources, and then dynamically update the bids using an insurance fund. For simplicity of evaluation and implementation we chose not to study this more advanced form of admission control here.

If strict admission control is implemented for all users only one guarantee level can be provided. To allow any number of guarantee levels, we strictly enforce only a portion of the allocation request, and make the remaining portion subject to statistical guarantees.

3   Simulations

In our simulations we study the price guarantees and dynamics, using varying levels of statistical and admission control guarantees offering multiple competing consumers service-level guarantees under different work-load situations.

The setup is as follows. A number of concurrent competing consumers submit jobs with inter-arrival-times (IAT) from an exponential distribution and performance requirements drawn from a normal distribution. The performance requirement is obtained from the number of work units that needs to be completed within a given deadline, and it translates to the share, q, of a resource that the consumer will bid for.

To simulate the fact that some users do not care about guarantees, but are only interested in best-effort service we designate a certain proportion of the work-load to be *best-effort* jobs. Those jobs are submitted by calculating the bid a consumer should spend based on the assumption that the price stays at the current mean value. This technically gives the guarantee, $g = 0.5$. All other jobs try to get a guarantee $g \geq 0.6$, and we then measure the guarantees obtained and the price paid under different levels of best-effort jobs. Each run of the simulated workload was configured with a single guaranteed service level, i.e. all jobs competing with best-effort jobs in a simulation run request the same guarantee level. We then measure and graph the average bid and obtained guarantee for a group of eigth subsequent jobs (based on completion time) requesting a certain guarantee level.

The guarantee obtained in a simulation run is calculated by measuring whether the current share of a job is greater than the required share each second that the job runs. The proportional share allocations are also recalculated each second. We configured the mean of the overall required shares to be higher than the available capacity in order to simulate resource contention and consumer competition.

The general simulation configuration is summarized in Table 1 and Table 2. #C is the number of consumers, #J is the number of jobs per consumer, t the deadline, and BE is the portion of best-effort jobs.

Table 1. General Configuration (All times in seconds)

#C	#J	q	IAT	g	t
4	32	$N(5.5/16, .25)$	$Exp(8)$	$(0.6, 0.9)$	16

Table 2. Individual Simulation Configuration

Simulation	BE	Strategy
I	0.75	statistical guarantee
II	0.25	statistical guarantee
III	0.25	admission control

3.1 Simulation I: 75% Best-Effort with Statistical Guarantees

In the first simulation we look at a work-load with a high portion of best-effort jobs (75%) that can make way for the smaller portions of jobs requiring guarantees. No admission control is used in this simulation, just statistical guarantees. In Figure 1,

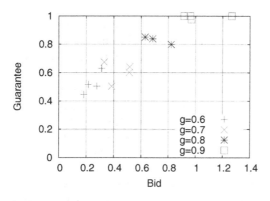

Fig. 1. Bids vs. obtained guarantees for statistical guarantees and 75% Best-Effort jobs

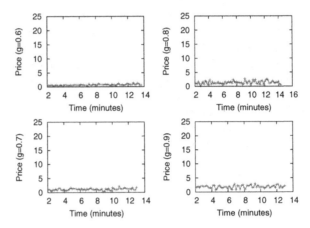

Fig. 2. Price over time for statistical guarantees and 75% Best-Effort jobs

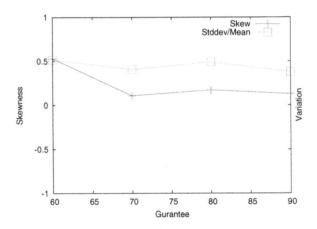

Fig. 3. Price skewness and variation for statistical guarantees and 75% Best-Effort jobs

where each marked point is an average of eight subsequently completing jobs, we see that there is a clear separation left to right and from bottom to top between the different guarantee levels. Jobs with higher guarantee requirements were bidding more (x-axis) and also obtained a higher guarantee (y-axis). This tells us that statistical guarantees worked well when giving consumers their guarantees in this scenario.

We also study the price dynamics. In Figure 2 we can see that the price is stationary although it has a high variance. Note that the first two minutes are not shown because this time is used to bootstrap the simulations. In Figure 3 the variation is high but stable, the skew is positive and varies between 0 and 0.5. A positive skew of the price distribution means that more jobs pay a higher price for a guarantee level than would normally (e.g. by Gaussian distribution models) be expected from the mean and the variance. Skewness can thus be viewed as an indication of how risky the computational market is [4].

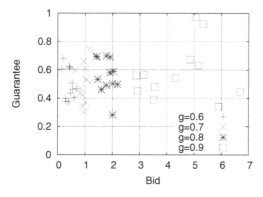

Fig. 4. Bids vs. obtained guarantees for statistical guarantees and 25% Best-Effort jobs

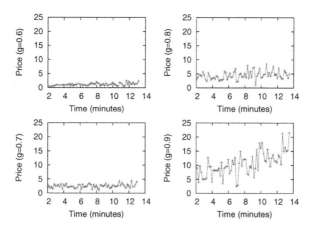

Fig. 5. Price over time for statistical guarantees and 25% Best-Effort jobs

3.2   Simulation II: 25% Best-Effort with Statistical Guarantees

We now decrease the portion of best-effort jobs to 25% and consequently the portion
of jobs requiring guarantees increases to 75%. In Figure 4 we can see that the guarantees obtained for the different guarantee-levels are seemingly randomly layered. The
higher bids and requested guarantees do not necessarily yield a higher obtained guarantee as before. This can be explained by the load being too high for the provider to offer
everyone the required guarantees.

Looking at the price fluctuations in Figure 5, there is a clear trend of inflation in
particular for $g = 0.9$ (bottom right). Also note that simply compensating for the bid
based on expected inflation would just accelerate this trend. In Figure 6 we see that
both the variance and the skewness of the price distribution exhibit similar behavior as
in Simulation I.

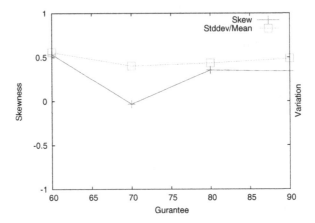

Fig. 6. Price skewness and variation for statistical guarantees and 25% Best-Effort jobs

3.3 Simulation III: 25% Best-Effort with Admission Control Guarantees

Finally we run a simulation with the same load configuration as in the previous simulation, i.e, 25%, best-effort jobs, but now we offer admission control for all non best-effort jobs. An admission control fee of $r = 0.05$ percent points and an enforcement portion of 30% was used. To simulate the important task of an admission control mechanism to allow users to defer their job submissions based on admission results, we defer and resubmit all guarantee jobs that cannot get at least 70% of their work load guaranteed. The time to wait before resubmission is determined randomly with a uniform distribution ranging $1 - 10$ seconds. In Figure 7 it is now again apparent that higher bids also give higher guarantees. Although the separation is not as clear as in Simulation I, it is clearly better than in Simulation II. The separation received is related to the proportion of the job that is strictly enforced. In the case of the entire job being strictly enforced all requested levels result in a 100% guarantee. If the enforcement proportion is made

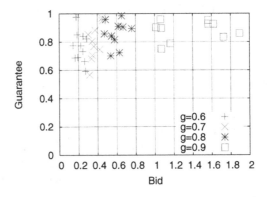

Fig. 7. Bids vs. obtained guarantees for admission control guarantees and 25% Best-Effort jobs

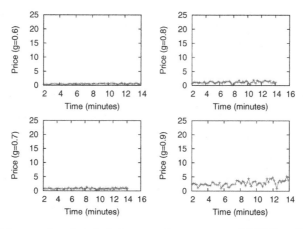

Fig. 8. Price over time for admission control guarantees and 25% Best-Effort jobs

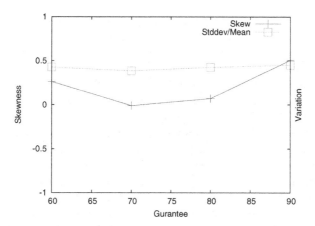

Fig. 9. Price skewness and variation for admission control guarantees and 25% Best-Effort jobs

too low, the reulsts will converge to those of Simulation II, that is, requested guarantees cannot be met reliably.

Figure 8 indicates that the inflation is now gone, and Figure 9 shows that the price distribution variation and skewness are similar to the previous two simulations. The penalty for the higher guarantees for some users rests partly on the best-effort jobs and partly on the fact that only a portion (70%) of the entire job run is strictly reserved. We should note that the overall load in this simulation is lower and thus the average bid for the jobs that are let through are obviously lower due to some jobs being refused to run by the admission control. The main point here, though, is that we can add admission control as a compromise between reservations and best-effort allocations in scenarios when statistical guarantees fail.

We summarize the results of the simulations in Table 3. Although the price distribution variation is the same in all the simulations, Simulation II exhibits a higher variance in guarantee levels delivered, in addition to not being able to deliver on the requested guarantee level.

Table 3. Summary of mean and variation of obtained guarantee levels when requesting 60, 70, 80, and 90% guarantees. All values are in percent.

Simulation	60		70		80		90	
	μ	σ/μ	μ	σ/μ	μ	σ/μ	μ	σ/μ
I	52	15	61	12	83	3	99	1
II	50	22	50	27	56	22	58	34
III	76	14	79	16	86	11	88	8

4   Related Work

There is a substantial body of work on Internet Protocol quality-of-service enforcement, represented by the two IETF specifications IntServ [5], and DiffServ [6]. The IntServ specification takes the approach of reserving paths for individual users, and thus does not scale as well as the DiffServ approach, which is based on marking individual packets with different *per-hop behaviors* in a stateless and decentralized architecture. Wang [7] gives an overview of lessons learned and the pros and cons of the reservation approach which can be implemented with IntServ versus the proportional share approach which can be built on top of DiffServ. The conclusion was that fixed allocations over a point-to-point path incur too much overhead for most of the web traffic, it is difficult to determine the resource requirements a priori, inter-ISP relationships make end-to-end reservations complicated, and traffic policing breaks down in the event of partial allocation failures. All of these factors result in many IP reservation providers over-provisioning their network capacity, leading to poor utilization. Wang therefore makes a case for a proportional share model [8] where each user receives a proportional share of the currently available bandwidth according to her contribution or spending. We are facing the same issues and trade-offs when allocating computational resources across large distributed systems. However, new virtualization technology and the fact that many of the resources are localized (e.g. CPU, memory, disk) makes it worth revisiting the reservation concepts.

One of the most critical parts of the IntServ architecture is the admission control component, and consequently there has been an extensive effort on designing efficient algorithms for deciding which packets are to be dropped versus served, and how routers and switches should be configured to shape the traffic according to the QoS levels promised to users. Knightly and Shroff provide an evaluation of the different admission control algorithms available for IP traffic shaping in [9]. The dilemma of denying access to flows that might have been served leading to underutilization compared to serving requests that will break existing QoS contracts makes it hard to use coarse statistical bounds and too simplified assumptions about traffic flow distributions. Put differently, both accuracy maximization and risk minimization are desired. The algorithms that accounted for economies of scale and not simply looked at the statistical properties of individual flows were shown to perform much better on average. Again, our admission control decision differs from the IP flow one, in that we can, through virtualization, more directly enforce that an admitted request stays within its bounds. Our decision is thus more about making sure that the provider does not lose out on utilization or profit by admitting low priority tasks prematurely.

MacKie-Mason et. al. [10] investigate how price predictors can improve users' bidding strategies in a market-based resource scheduling scenario. Their conclusion is that even very simple predictors, such as taking the average of the previous round of auctions, help improving expected bidder performance. Another interesting result is that the main reason the predictor strategies outperform memory-less strategies is the fact that the binary decision of whether to participate in an auction can save the bidder more money than accurately estimating exactly how much to bid to obtain a certain performance level. Although, the high-level goal of this work is strikingly similar to ours they investigate a very different allocation and auction scenario, where combinatorial preferences exist and there is a risk of only receiving subsets of the preferred resources. Furthermore, first price winner-takes-it-all auctions are employed, as opposed to proportional share auctions in our work. Nevertheless, their results are encouraging. Another successful use of economic predictions to optimize bidding strategies is described by Wellman et. al. in [11], where bidding agents determine their bids and auctions to enter based on the expected market clearing price in a competitive or Walrasian equilibrium. To find this price they employ the process of *tatonnement* which involves determining users' inclination to bid a certain value given a price-level. Wellman et. al. compare their competitive analysis predictor to simple historical averaging and machine learning models as employed in the Trading Agent Competition (TAC) and conclude that strategies not only considering background history data but also instance-specific data in the predictions provided a competitive advantage. Finally, their competitive predictor performed on-par with the best machine learning predictor. The conditional probability of price dynamics given a certain price-level would be very useful to collect in our case too to get a full picture of the usage pattern. However, in large-scale systems with users entering and leaving the market at will, and large real-valued price ranges it quickly becomes impractical for our purposes, so we assume this behavior is incorporated in the price history itself.

5   Conclusions

We have studied the effects of bidding for virtualized resource shares using price predictions and admission control. For the predictions to be effective there must either be a sufficiently large portion of best-effort bidders, who can decrease their shares when there is contention, or an admission control mechanism refusing access to requests that would break the existing QoS contracts.

Whether a consumer should spend extra money on getting a higher level of guarantee through an admission control contract, thus depends on the contention among consumers requiring high guarantees. Price history and price distribution analysis serve as good indicators for determining whether this is the case. Conversely, providers would be interested in knowing how to partition their resources between the admission control market versus the best effort market depending on the price fluctuation characteristics and usage pattern.

Future work includes reproducing the simulation results in experiments in a live Grid market deployment (presented in [2]), more in-depth analysis of how providers can dynamically partition their resources for contract markets, and adding more sophisticated option and risk-hedging reservations to the admission control mechanism presented here.

Acknowledgments

We thank our colleagues Bernardo Huberman, Li Zhang, Fang Wu, Ali Ghodsi and Scott Clearwater for fruitful discussions. This work would not have been possible without the funding from the HP/Intel Joint Innovation Program (JIP), our JIP liason, Rick McGeer, and our collaborators at Intel, Rob Knauerhase and Jeff Sedayao.

References

1. Sandholm, T., Lai, K.: Evaluating Demand Prediction Techniques for Computational Markets. In: GECON '06: Proceedings of the 3rd International Workshop on Grid Economics and Business Models, Singapore (2006)
2. Sandholm, T., Lai, K., Andrade, J., Odeberg, J.: Market-based resource allocation using price prediction in a high performance computing grid for scientific applications. In: HPDC '06: Proceedings of the 15th IEEE International Symposium on High Performance Distributed Computing, pp. 132–143 (2006)
3. Lai, K., Rasmusson, L., Adar, E., Sorkin, S., Zhang, L., Huberman, B.A.: Tycoon: an Implemention of a Distributed Market-Based Resource Allocation System. Technical Report arXiv:cs.DC/0412038, HP Labs, Palo Alto, CA (2004)
4. Mandelbrot, B., Hudson, R.L.: The (Mis)behavior of Markets: A Fractal View of Risk, Ruin, and Reward. Basic Books, New York (2004)
5. Braden, R., Clark, S., Shenker, S.: Integrated services in the internet architecture. RFC 1633, IETF (1994)
6. Blake, S., Black, D., Carlson, M., Davies, E., Wang, Z., Weiss, W.: An architecture for differentiated services. RFC 2475, IETF (1998)
7. Wang, Z.: A case for proportional fair sharing. In: IWQoS '98: Proceedings of the Sixth International Workshop on Quality of Service, pp. 33–35. IEEE, Los Alamitos (1998)
8. Wang, Z.: Usd: Scalable bandwidth allocation for the internet. In: HPN, 351–361 (1998)
9. Knightly, E.W., Shroff, N.: Admission control for statistical qos: Theory and practice. 13(2), 20–29 (1999)
10. MacKie-Mason, J.K., Osepayshvili, A., Reeves, D.M., Wellman, M.P.: Price prediction strategies for market-based scheduling. In: ICAPS, 244–252 (2004)
11. Wellman, M.P., Reeves, D.M., Lochner, K.M., Vorobeychik, Y.: Price prediction in a trading agent competition. J. Artif. Intell. Res (JAIR) 21, 19–36 (2004)

DFCA: A Flexible Refundable Auction for Limited Capacity Suppliers

Zhixing Huang[1,2] and Shigeo Matsubara[2]

[1] Semantic Grid Laboratory, Faculty of Computer and Information Science
Southwest University, Chongqing 400715, China
huangzx@swu.edu.cn
[2] Global Information Network Laboratory, Department of Social Informatics
Kyoto University, Kyoto 606-8501, Japan
matsubara@i.kyoto-u.ac.jp

Abstract. This paper proposes a novel auction-based mechanisms named Decreasing Cancellation Fee Auction (DCFA) for task allocation in the environment where a service provider has finite capacities and consumers could withdraw their bids. We consider a new type of auction called the refundable auction, i. e. refund means that a consumer's showing up is uncertain and he can get back partial of his payment if his cancellation or no-show occurs. This mechanism can boost seller revenue, satisfy incentive compatibility, individual rationality and still hold a high efficiency.

Keywords: Refundable auction, price matching, incentive compatibility, VCG mechanism, advance reservation.

1 Introduction

Due to the geographic distribution of resources that are often owned by different organizations with different usage policies and cost models, and varying loads and availability patterns, the task of resource management and scheduling in these environments is a complex undertaking. *Distributed Computational Economy* [3] has been recognized as an effective metaphor for the problem of such management. In particular, auction has been recognized as an effective method for the management of distributed resources [3,6], because it enables the regulation of supply and demand for resources, provides economic incentive for service providers, and motivates the service consumers to trade off among deadline, budget, and the required level of quality-of-service. Typical applications include task assignment, distributed scheduling, etc.

The distributed system has a highly dynamic environment [21] with servers coming on-line, going off-line, and with continuously varying demands from the clients. Therefore, the function of *Advance Reservation* has been strongly recommended into supporting the allocation and scheduling mechanisms, because the computing resources are usually not storable and the capacity available today cannot be put aside for future use [6]. It should be noted that the auction-based advance reservations are currently being added to some economic-based toolkits,

D.J. Veit and J. Altmann (Eds.): GECON 2007, LNCS 4685, pp. 83–97, 2007.

such as GridSim [1] which has integrated First-Price Sealed-Bid, English, Dutch and Continuous Double auction protocols. However, there may exist many uncertainties of consumers' requests (bids). The possibilities for breaks in actions include [9]: an erroneous initial valuation or bid, unexpected events outside the winning bidder's control, information obtained or events that occurred after the auction, etc. For instance, in Data Mining applications, users may cancel the visualization step when the result is not interesting enough or the mining procedure could not be fulfilled. Hence, an importance feature, as noted in GRAAP-WG[1] the advance reservation protocol should allow consumers to cancel or alter their booked services.

In economics-based allocation methods, *refund policies* are used to control for the selection of potential customers who make reservations but differ with respect to their cancellation probabilities. Refund policy assumes that a consumer pays for the service during the reservation is made, but the consumer gets partial (or all and no) refund when his cancellation or no-show occurs. Refunds are widely observed in almost all privately-provided services and also to some degree in retail in industries. Most noticeably, refunds are heavily used by airline companies. Refundable bookings tend to attract consumers who are likely to cancel or not show up for the service, and deter consumers who are less likely to cancel and are therefore more price sensitive [20]. However, the cancellation and refund issue in auctions has been discussed little in both economics and computer science literatures.

In this paper, we concentrate on the design of *partially refundable auction* mechanism of advance reservation systems in which consumers' show-ups are uncertain and their personal information are unknown by the service providers. We demonstrate the nonexistence of any mechanism which satisfies seller's profit-maximizing, individual rational and incentive compatibility simultaneously. We propose a flexible auction mechanism that can satisfy incentive compatibility, individual rationality, and still obtain a high efficiency. The remainder of this paper is organized as follows. In Section 2, we present the main related work in this area. In Section 3, the basic model of the refund auction is described. We point out that there does not exist any uniform pricing and cancellation fee allocation mechanism that could maximize seller's profit and satisfy incentive compatible at the same time. In Section 4, we present Decreasing Cancellation Fee Auction mechanism. In Section 5, the experimental comparison of our mechanisms with an ideal optimal algorithm and other counterparts are evaluated. Finally, we conclude this paper and discuss future work in Section 6.

2   Related Work

Incentive mechanism design is an important issue not only in Economics but also recently in E-commerce and Distributed Artificial Intelligence, obviously

---

[1] Advance Reservation State of the Art: Grid Resource Allocation Agreement Protocol Working Group,
http://www.fz-juelich.de/zam/RD/coop/ggf/graap/sched-graap-2.0.html

this issue should not been ignored in Grid Economics. Incentive compatibility means the dominant strategy for a bidder is to bid to his true valuation thus bidders'decision-making could be simplified even in the highly complex trading environment [11,12]. The best-known such payment rule is the Vickrey-Clarke-Groves (VCG) mechanism [7,8,23].

Uncertainty is quite common in computer environment for the unexpectable nature of the future events, it makes incentive mechanism design more complex. One long-standing problem in mechanism design is to design the optimal multidimensional incentive mechanism [28]. Porter et al. introduce the notion of fault tolerant mechanism design, they have proved an impossibility result that there does not exist a mechanism that satisfies incentive compatibility, individual rationality and social efficiency when dependencies exist between tasks [16]. Holland and Sullivan present the non-existence of mechanism to incentivize truthful bidding when robust allocations are required in a revocable combinatorial auction [10], however they assume that bid-taker knows bidders' private showup probabilities. Hurberman et al. propose a pricing mechanism induces truth-telling on the part of users reserving the service, one limitation of their work is that they assume all the users have the same valuation about the service [2,13,24], but users' valuation should be different in the real world.

Refunds are wildly observed in almost all privately-provided services which alleviates the consumers' risk of *uncertainty* and also could increase seller's revenue. However, the refundable auction has not been well discussed in both economic and AI literatures. The refundable auction problem are complicated by *incentive compatibility* constraint, multidimensional consumer type and private cancellation uncertainty. Without the cancellation constraint, the incentive compatibility could be guaranteed by strategy-proof protocols such as Vickrey-Clarke-Groves (VCG) mechanism and Price-Oriented Rationing-Free (PORF) mechanism [26]. If the private information could be reduced to one dimension, we could realize seller's profit-maximizing using the method proposed in [5]. If the uncertainty of the future event is public information, we could enumerate all possible allocation plans as proposed in [14] or uses only the part of the bid as the winner scoring rule to find the optimal allocation [4].

This work is greatly inspired by the work of Ringbom and Shy [17,20]. Shy gives a compressive introduction about advance booking and various refund strategies in his upcoming book [20]. They developed methods for calculating optimal rates of partial refunds for profit-maximizing and social surplus maximizing when price of the goods, but their model assumes that price is exogenously given and auction-based mechanism has not been discussed in their current works. We will show in the following, using auction-based mechanism could greatly increase the profit of the service providers.

3   Problem

Suppose a seller (or service provider) has m unit perishable goods (or service) and there are n $(n > m)$, consumers are willing to buy only one unit of the good. Consumer i's valuation about a unit of the good is β_i, where $\beta_i \in [0, 1]$. Let $\theta_i \in [0, 1]$ denote the probability that consumer i will show up and consume the service at contracted delivery time. We call θ_i the probability of showing up. Therefore, the probability of a cancellation is $1 - \theta_i$. We assume that the valuation and the probability of showing up are exogenous variables, thus these variables cannot be changed by consumers. Let p denote the price of the service and r denote the *cancellation fee* when the reservation is cancelled by the consumer or the consumer does not show up.[2] We assume that $p \geq r$. We will not distinguish between a cancellation and a non-show-up in this paper. The payoff of the confirmed consumer i as follows:

$$c_i = \begin{cases} p & \text{if show up,} \\ r & \text{otherwise.} \end{cases}$$

The expected utility of the confirmed consumer i is: $u_i = \theta_i(\beta_i - p) - (1 - \theta_i)r$. In this refundable pricing mechanism, consumer i needs pay p if his request has been booked and has been completed, but he needs to pay r if the booked request has been cancelled by himself, and need not pay if the request has been declined. In other words, we assume that all consumers must pay for the service during the reservation is made, so consumer get $p - r$ refund when his cancellation occurs. The expected revenue of the seller from the confirmed consumer i is: $\pi_i = \theta_i p + (1 - \theta_i)r$. To maximize the expected profit, the seller will select m most profitable consumers.

Theorem 1. *There does not exist a uniform pricing and cancellation fee allocation mechanism that could satisfy seller's profit-maximizing and incentive compatible simultaneously.*

Proof (Sketch). If the service provider knows all the private information of bidders, the optimal pair (p, r) which maximize the service provider's expected revenue could be calculated as solving following optimal problem: $\max_{p,r} \sum_{i=1}^{N} x_i \cdot [p \cdot \theta_i + (1 - \theta_i) \cdot r]$, s.t.$0 \leq r \leq p \leq 1, \sum_i^n x_i \leq m, x_i \theta_i \cdot (p - \beta_i)/(1 - \theta_i) \leq r, x_i \in \{0, 1\}$. However, consumers could fake bids with different combinations which may cause a loss of the service provider's revenue, the detailed example is as described below.

Example 1 (fake bid by loser). Suppose there are four consumers and two unit goods, bidders' valuations and showup probabilities are shown in Table 1(a). If consumers truthfully report their private information, the optimal price and cancellation fee pair should be $(0.45, 0.45)$, which means that bidders 1 and 2 are winners, in that time $u_1 > 0$, $u_2 = 0$, $u_3 < 0$, $u_4 < 0$, and the seller's expected revenue is 0.9. However, if bidder 3 overstates his showup probability from 0.2 to 0.5 while understates his valuation from 0.9 to 0.8, as shown is Table 1(b),

---

[2] Cancellation fee r can also be interpreted as *advance payment* as the economic model discussed in [2,13,24]. The advance payment assume that the customers pay r before they consume the goods and they pay $p - r$ after their consumptions.

Table 1. Example 1

		(a)						(b)			
	b1	b2	b3	b4			b1	b2	b3	b4
β	0.8	0.5	0.9	0.3		β	0.8	0.5	0.8	0.3
θ	0.7	0.9	0.2	0.4		θ	0.7	0.9	0.5	0.4

then bidders 1 and 3 will win the goods. Based on those fake information, the optimal price and cancellation fee pair would be $(0.8, 0)$, in that time $u_1 > 0$, $u'_3 = 0 (u_3 > 0)$, $u_2 < 0$, $u_4 < 0$. In such a case, the seller predicts its expected revenue is 0.96, in contrast, his true expected revenue is $0.56 + 0.16 = 0.72$, which is less than 0.9. The seller will suffer 20% loss on his expected revenue when the bidder 3 fakes his bid.

Example 2 (fake bid by winner). As in Table 1(a), bidder 1 and 2 also have the incentive to lie. For instance, if they fake their bids as $\beta = 0.4, \theta = 0.5$, each of them can still obtain one unit of goods. In this time, the calculated optimal price and cancellation fee pair is $(0.23, 0.17)$. However, the seller's true expected revenue is 0.436 that is less than half of the ideal value.

Theorem 1 also means that if we cannot guarantee incentive compatibility, the seller's profit-maximizing cannot be guaranteed. The difficulty in designing the optimal incentive compatible auction is that the bids and types are both multi-dimensional, especially with payment uncertainty at the same time [28].

The incentive compatibility constraint can be easily satisfied by extending the standard VCG mechanism. The rules of the extended VCG mechanism are described as follows: 1) all bidders are required to report their expected valuation $\gamma_i = \theta_i \beta_i$ (may not truthful); 2) the closing price p is the $(m+1)$th highest γ_{m+1}; 3) withdrawal after the winner determination phase is not allowed, in another words, it is a no-refundable in this mechanism, which means that $r = \gamma_{m+1}$.

Definition 1. *We define the expected social welfare as the sum of the revenue of service provider and the utility of bidders:*

$$W = \sum_{i \in \Lambda} \left(\theta_i p + (1 - \theta_i) r \right) + \sum_{i \in \Lambda} \left(\theta_i (\beta_i - p) - (1 - \theta_i) r \right) = \sum_{i \in \Lambda} \theta_i \beta_i \quad (1)$$

where Λ is the set of winners.

The first item of Equation (1) indicates the expected revenue of the service provider, and the second item indicates the sum of the winners' utilities. We also simply call expected social welfare as efficiency in the later.

Theorem 2. *The extended VCG mechanism satisfies individual rationality, incentive compatibility and social efficiency.*

Proof. Suppose γ_{m+1} is the *(m+1)*th highest bid, the utility of bidder i is: $U_i = \theta_i(\beta_i - \gamma_{m+1}) - (1 - \theta_i)\gamma_{m+1} = \theta_i\beta_i - \gamma_{m+1} = \gamma_i - \gamma_{m+1}$. From this equation, we can see that the original two dimensional problem now is reduced to one dimensional problem. It is clear that: if bidder i overstates it will cause a non-positive utility when $\gamma_i \leq \gamma_{m+1}$; Meanwhile when $\gamma_i > \gamma_{m+1}$, over-reporting is useless, i.e. it could not increase bidder's utility. Thus, this mechanism satisfy incentive compatibility. For each bidder truthful telling his private information is a dominant strategy and the price is the *(m+1)*th highest bid γ_{m+1}, the individual rationality of this mechanism is immediate. Since the winner of the auction are the m bidders whose bid $\theta_i\beta_i > \gamma_{m+1}$, the expected social welfare as defined in Equation 1 is maximized.

However, the extended VCG mechanism could not guarantee to obtain a sufficient high revenue. For instance, when we apply the extended VCG mechanism given data in Table 1(a), the seller's profit is only $0.1800 \times 2 = 0.3600$, which is far lower than optimal value.

4   Decreasing Cancellation Fee Auction

In the following section, we first discuss the Fixed Cancellation Fee Auction (FCFA) mechanism before we propose the Decreasing Cancellation Fee Auction (DCFA) mechanism. The DCFA can boost seller revenue, satisfy incentive compatibility, individual rationality and still hold a high efficiency.

4.1   Fixed Cancellation Fee Auction

The detailed FCFA protocol is described as below. Firstly, the seller announces a posted cancellation fee r before the auction begins. Then based on the posted cancellation fee r, each bidder reports his maximum acceptable price $p_i = \beta_i - r(1 - \theta_i)/\theta_i$. After all the bidders submitted their bids, the seller selects the m consumers with the most highest price as the winners of the auction. The payment rules are described below: i) If there are more than m consumers, the winners' payment will be determined by the *(m+1)*th highest bid, which is the highest of all losing bids. ii) When there are less than m consumers, all the consumers are winners, the payment is r, the lowest price of the seller to provide the service. iii) As mentioned before, in both cases, if a confirmed consumer cancels its reservation or does not show up at the deliver time, the consumer's payoff is r. First of all, the most important task for the resource provider, in this protocol, is to set an optimal cancellation fee r^* before the auction begins. The algorithm to calculation r^* is discussed in Appendix A.

Theorem 3. *The dominant strategy of bidder i under Fixed Cancellation Fee mechanism is he truthfully reports his private information p_i.*

Proof. The Fixed Cancellation Fee mechanism has simplified the original multidimensional problem to one dimensional problem: bidder's private information is indirectly revealed in its maximum acceptable price p_i. The Fixed Cancellation Fee mechanism uses $(m+1)$th-price auction protocol for the winner determination. As discussed in [25], the $(m+1)$th-price sealed-bid auction is incentive compatible for single-unit buyers under the independent private values model. If he reports lower price $p'_i < p_i$, he may face the risk of losing the auction, and it is also useless since he cannot manipulate the closing price by understating p_i. On the other hand, if the bidder reports higher price $p'_i > p_i$, if $p > p_i$ he will obtain a negative utility, and over-reporting is also useless, because its payment p is highest loss price $\hat{p}_{m+1}$ when $k > m$ or cancellation fee r when $k \leq m$. So this mechanism satisfies the incentive compatibility property: a bidder truthfully reports its private information p_i is a *dominant strategy*.

One limitation of the FCFA mechanism is that the number of the qualified bidders may be less than the total number of the goods, this will leave some capacity unutilized thereby resulting in a loss to service providers. Now, in this section, we will extend the FCFA mechanism to the flexible DCFA mechanism. We also assume that the service provider has m units perishable goods (or service). The auction is proceeded iteratively according to a series of the non-negative cancellation fees in the decreasing order $r_1 > r_2 > \cdots > r_l > 0$. Meanwhile, the winners of the auction are selected sequentially based on their bids. The *price matching* method [22] is adopted in this method, i.e. the final cancellation fee and the price is the lowest valuation in all those rounds, thus the bidders need not worry about the possible loss while the cancellation fee and price are decreasing in later rounds.

4.2 Allocation Protocol

The auctions are proceeded in rounds, suppose there are l rounds, and each round holds a Fixed Cancellation Fee auction. The auction rules are described as follows:

1. In round of the auction j, the price p_j must be no less than the cancellation fee r_j, i.e. $p_j \geq r_j$. Intuitively, it will be unrealistic in the real market that the mechanism with cancellation fee is more than the price of the goods.
2. Let m_j represents the number of units available in round j, the number of the valid bids, where $b_i \geq r_j$, is w_j. If $m_j + 1 \leq w_j$, then the bidders with highest m_j bids will be the winners of the auction, the price will be the (m_j+1)th highest bids among w_j bids. If $m_j \geq w_j$ and $w_j \geq 2$ then the $w_j - 1$ highest bidders are winners, the price will be the w_j-th highest bid, i.e. the lowest valid bid in round j. In another words, we perform the $(m+1)$th price auction in each round, no matter whether the valid bids are more or less than the number of goods.
3. However, if there is only one bidder in the auction, he could manipulate the price, thus if there is only one bidder in the current round, the auction turns to the next round. It means that the valid bid number in each round should be greater than two, i.e. $w_j \geq 2$, $j \in [1, l]$.

4. All the winners reserve a unit of good with price $p = \min_{j \in [1,J]} \{p_j\}$ and cancellation fee r_J, where J is the round index that the last unit is sold or the last round $(J = l)$ when there is still has unsold but with no valid bidders.

The concrete procedure is described in Algorithm 1, where parameter m_j represents the number of units available in round j, and w_j represents the number of the valid bids in round j. The algorithm includes four steps: Step 1 executes *initiation* process (lines 1 to 2) that initializes all the parameters; Step 2 is the *booking* process (lines 3 to 21), which accepts the bids and selects the winners into queue Q; Step 3 is *price matching* step (lines 22 to 24), this step matches all the winners' booking prices and cancellation fees with the lowest ones; In addition, the final step, the *overbooking* step is the optional step and will be discussed in our future work.

Algorithm 1: Decreasing Cancellation Fee Auction
```
1: initiate the cancellation fees r[1], r[2],..., r[l];
        /* where r[1]> r[2]>...> r[l]> 0.*/
2: j:= 1; m[1]:= m; p:= 1;
3: while (m[j]>= 0 and j<= l) do
4:      execute the FCFA with cancellation fee r[j];
5:      w[j]:= the number of bids which satisfies b[i]>= r[j];
6:      if (w[j]<= m[j] and w[j]>= 2) then
7:         append the highest w[j] bids in the winner queue Q;
8:         m[j+1]:= m[j]-w[j]+1;
9:      end if
10:     if (w[j]>= m[j]+1 and w[j]>= 2) then
11:        append the highest m[j] bids in the winner queue Q;
12:        m[j+1]:= 0;
13:     end if
14:     if (w[j]>= 2) then
15:        if (p[j]< p) then
16:            p:= p[j];
17:        end if
18:        r:= r[j];
19:     end if
20:     j:= j+1;
21: end while
22: for all bids in Q do
23:     p[i]:= p; r[i]:= r;
24: end for
25: overbooking (optional)
```

Theorem 4. *The DCFA stratifies Individual Rationality and Incentive Compatibility properties.*

Proof. The Decreasing Cancellation Fee auction protocol is an variance of Fixed Cancellation Fee Auction protocol, the allocation of goods are separated in multiple rounds. For the cancellation fee and the price of goods are non-increasing in the auction, the individual rational is satisfied. According to the auction rules, the final cancellation fee is based on the round of latest winner in the action, and the price is the lowest price in all those rounds, thus the bidders need not worry about the possible lose while the cancellation fee and price are decreasing in later rounds. In each round, the *(m+1)*th price auction is held in each round, the price could not be manipulated by any bidder. In addition, there is more riskily to be lost of the auction in later rounds. So the bidder's best strategy is to bid as earlier as possible, That is, selecting the round of the auction is useless and risky and thus incentive compatibility property is guaranteed. *Best-response* strategy is bidder's dominant strategy.

Example 3. Now, we give an example to demonstrate the process of this protocol, suppose the bidders' information as shown in Table 1(a), and the cancellation fee series rs are $\{0.35, 0.25, 0.15, 0.05\}$. Auction begin with $r_1 = 0.35$, bidder 1 and 2 bid with price 0.65 and 0.46, according to rule 2, bidder 1 reserve one unit of goods, with the deferred price 0.46, then auction goto the next round $r_2 = 0.25$. For this round, only bidder 2 bids, according to rule 3, the auction move to the next round. When $r_3 = 0.15$, bidder 2, 3, 4 bid with price $0.48, 0.3, 0.075,$, the last unit good goes to bidder 2 with price 0.3. The final allocation result is bidder 1 and 2 each win one unit of goods, with $p = \min\{0.45, 0.3\} = 0.3$ and $r = 0.15$. The expected revenue of the sell is $0.255 + 0.285 = 0.56$, which is higher than that of the extended VCG mechanism and the result of *Example 2*.

This mechanism could be easily extended to support multiple unit demand and still keep incentive compatible property through adapting the similar method proposed in [27], when marginal values of all participates decrease or remain the same. The detailed method is discussed in Appendix B.

5   Experiment and Comparison

We compare our propose mechanism with the ideal situation and the fixed price-refund pair method. In the ideal situation, we assume that the seller knows all the information about bidders, although it is hard to induce all those information as discussed in previous section. An other method, the fixed price-cancellation fee pair method [3], could be calculated based on the distribution of the distribution of users' type. The optimal fixed pair (p, r) could also be calculated by iteratively using the algorithm proposed in [17], they calculate the optimal refund level when price is exogenously given.

5.1   Experimental Setting

In each experimental setting, the bidder's valuation and showup probability are uniformly and independently distributed the interval $[0, 1]$, the minimal

---

[3] For refund rate is equal to $p - r$, we also called this method fixed price-refund pair method or posted (p, r) pair in short.

Table 2. Comparison of average expected revenue, efficiency, users' utility, price and cancellation fee in different refundable auctions when $(m, n) = (3, 20)$

Mechanism	Revenue	Efficiency	Users' Utility	Price	CF
Decreasing Cancellation Fee	1.6109	1.7396	0.1287	0.5761	0.4170
Fixed Cancellation Fee	1.4888	1.8834	0.4091	0.5470	0.3000
Posted (p, r) Pair	1.3069	1.6030	0.2961	0.6200	0.2100
Optimal	1.7648	1.8782	0.1135	0.7254	0.2139
Extended VCG	1.3965	1.8972	0.5007	0.4655	0.4655

Table 3. Comparison of average expected revenue, efficiency, users' utility, price and cancellation fee in different refundable auctions when $(m, n) = (3, 40)$

Mechanism	Revenue	Efficiency	Users' Utility	Price	CF
Decreasing Cancellation Fee	1.9487	2.0569	0.1083	0.6700	0.5532
Fixed Cancellation Fee	1.8522	2.1844	0.3322	0.6502	0.4300
Posted (p, r) Pair	1.6697	1.9072	0.2375	0.7200	0.2100
Optimal	2.0888	2.1827	0.0938	0.7907	0.2578
Extended VCG	1.8251	2.1999	0.3748	0.6084	0.6084

increment of two random value is 0.01. In each same setting, the auction will be run at least 1000 times.

5.2   Comparison of the Expected Revenue

In this experiment, we compare the average expected revenue, efficiency, users' utility, price and cancellation fee (CF) among different mechanisms. Due to the space limitation, we only draw two experimental results when $(m, n) = (3, 20)$ and $(3, 40)$ to demonstrate the relative performance of those the methods since we get similar result in the other situations. The underlined number in the tables indicates the value is fixed. More specially, we set rs are $\{0.7, 0.6, \cdots, 0.1\}$ in the DCFA method. From Table 2 and 3, we can see that FCFA and DCFA mechanisms can obtain more profit than fixed price-refund pair and extended VCG method. In addition, DFCF can approach to the optimal revenue result as the bidder number increases.

Compared with ideal optimal method which seller knows all bidders' private information, the inefficient of FCFA is mainly because that fixed cancellation fee restrained the bidders' participate the resource competition. Especially, in some cases, it causes the goods could not be totally sold out. For example, in fixed cancellation fee method simulation $(m, n) = (3, 20)$, there are nearly 10% runs that the number of valid bidders are less than the resource number. It causes the major loss of the sellers' revenue. DFCF overcomes this shortcoming by sequentially adjusting cancellation fees.

DFCF auction is insensitive to the selection of r series, but the value of r_1 should at least above r^* that is calculated in FCFA. The experiment results are shown in Table 4(a). On the other hand, as shown in Table 4(b), it is clear

Table 4. The selection of different r series

<table>
<tr><td colspan="5" align="center">(a)</td><td colspan="5" align="center">(b)</td></tr>
<tr><td>r_1</td><td>Interval</td><td>Number</td><td>Revenue</td><td>Efficiency</td><td>r_1</td><td>Interval</td><td>Number</td><td>Revenue</td><td>Efficiency</td></tr>
<tr><td>0.9</td><td>0.1</td><td>9</td><td>1.61</td><td>1.74</td><td>0.7</td><td>0.0125</td><td>56</td><td>1.72</td><td>1.76</td></tr>
<tr><td>0.8</td><td>0.1</td><td>8</td><td>1.61</td><td>1.74</td><td>0.7</td><td>0.025</td><td>28</td><td>1.69</td><td>1.76</td></tr>
<tr><td>0.7</td><td>0.1</td><td>7</td><td>1.61</td><td>1.74</td><td>0.7</td><td>0.5</td><td>14</td><td>1.66</td><td>1.76</td></tr>
<tr><td>0.6</td><td>0.1</td><td>6</td><td>1.61</td><td>1.74</td><td>0.7</td><td>0.075</td><td>9</td><td>1.64</td><td>1.75</td></tr>
<tr><td>0.5</td><td>0.1</td><td>5</td><td>1.60</td><td>1.72</td><td>0.7</td><td>0.1</td><td>7</td><td>1.61</td><td>1.74</td></tr>
</table>

that the bigger r's number the more revenue will be. The seller could trade off the auction lasting time and the revenue through select different r series. Although the implementation of auction-based methods may be more complex than posted-price based methods, auction-based methods could be more flexible and make more profit than the posted-price method and the extended VCG mechanism.

6   Conclusion and Future Work

In this paper, we propose a novel auction-based mechanism for task allocation in environments where service provider has finite capacities and consumers could withdraw their bids. We consider a new type of auction in which winner could withdraw We demonstrate that it is difficult to design an optimal auction protocol that satisfies profit-maximizing and incentive compatibility simultaneously. We explore two auction-based refundable mechanisms for boosting seller's revenue from the single stage and multistage perspectives. These mechanisms can satisfy incentive compatible and individual rational properties. The experimental results illustrate that these methods achieve higher revenue than the counterparts such as fixed price-refund pair method and extended VCG mechanism.

The Decreasing Cancellation Fee Auction mechanism can be easily extended support overbooking in the form of Leveled Commitment Contract [18,19]. We wish to further investigate how to negotiate optimally, or at least fairly, the sequential Leveled Commitment Contracts with different bidders. It is still an open question that how an service provider should allocate its scarce computational resources when evaluating different Leveled Commitment Contracts. Furthermore it will be interesting to extend these mechanisms to deal with the uncertainty in more complex auction protocols, such as double auction and combinatorial auction.

Acknowledgements. This work is supported by National Basic Research Program of China (973 project no. 2003CB317001). We are grateful to Prof. Toru Ishida, Dr. Xudong Luo and Dr. Yichuan Jiang for their insightful comments.

References

1. Assuncao, M.D., Buyya, R.: An Evaluation of Communication Demand of Auction Protocols in Grid Environments. In: Proceedings of the 3rd International Workshop on Grid Economics & Business, World Scientific Press, Singapore (2006)
2. Byde, A.: Incentive-Compatibility, Individual-Rationality and Fairness for Quality of Service Claims, Technique Reports, HPL-2006-27 (February 27, 2006)
3. Buyya, R., Abramson, D., Giddy, J., Stockinger, H.: Economic Models for Resource Management and Scheduling in Grid Computing. Journal of Concurrency and Computation 14(13-15), 1507–1542 (2002)
4. Chao, H.P., Wilson, R.: Multi-Dimensional Procurement Auctions for Power Reservation: Robust Incentvie-Compatible Scoring and Settlement Rules. Journal of Regulatory Economics 22(2), 161–183 (2002)
5. Che, Y.K.: Design Compition through Multidimensional Auctions. RAND Journal of Economics 24(4), 668–680 (1993)
6. Cheloitis, G., Kneyon, C., Buyya, R.: 10 Lessons from Finance for Commercial Sharing of It Resources, Peer-to-Peer Computing: The Evolution of a Disruptive Technology, 244–264 (2004)
7. Clarke, E.H.: Multipart Pricing of Public Goods. Public Choice 11, 17–33 (1971)
8. Groves, T.: Incentives in Teams. Econometrica 41, 617–631 (1973)
9. Harstad, R.M., Rothkopf, M.H.: Withdrawable Bids ss Winner'S Course Insurance. Operations Research 43(6), 334–339 (1998)
10. Holland, A., Sullivan, B.O.: Truthful Risk-Managed Combinatorial Auctions. In: Proceedings of IJCAI-07, Hyderabad, India (January 2007)
11. Huang, Z., Qiu, Y.: Resource Trading using Cognitive Agents: A Hybrid Perspective and Its Simulation. Future Generation Computer Systems 23(7), 837–845 (2007)
12. Huang, Z., Qiu, Y.: A Comparison of Advance Resource Reservation Bidding Strategies in Sequential Ascending Auctions. In: Zhang, Y., Tanaka, K., Yu, J.X., Wang, S., Li, M. (eds.) APWeb 2005. LNCS, vol. 3399, pp. 742–752. Springer, Heidelberg (2005)
13. Huberman, B.A., Wu, F., Zhang, L.: Ensuring Trust in One Time Exchange: Solving the QoS Problem. Netnomics 22(2), 161–183 (2006)
14. Matsubara, S.: Auction in Dynamic Environments: Incorporating The Cost Caused By Re-Allocation. In: Proceedings of the Fourth Autonomous Agents and Multi-Agent Systems (AAMAS-2005), pp. 643–649 (2005)
15. Menezes, F.M., Monteiro, P.K.: An Introduction to Auction Theory. Oxford University Press, Oxford (2005)
16. Porter, R., Ronen, A., Shoham, Y., Tennenholtz, M.: Mechanism Design with Execution Uncertainty. In: Proceedings of the 18th Conference in Uncertainty in Artificial Intelligence, pp. 414–421 (2002)
17. Ringbom, S., Shy, O.: Advance Booking, Cancellations, and Partial Refunds. Economics Bulletin 13(1), 1–7 (2004)
18. Sandholm, T., Conitzer, V.: Leveled Commitment Contracting: a Backtracking Instrument for Multiagent Systems. AI Magazine 23(3), 89–100 (2002)
19. Sandholm, T., Zhou, Y.: Surplus Equivalence of Leveled Commitment Contracts. Artificial Intelligence 142, 239–264 (2000)
20. Shy, O.: How To Price: a Guide to Pricing Techniques and Yield Management. Cambridge University Press, Cambridge (to appear, 2008)

21. Smith, W., Forster, I., Taylor, V.: Scheduling with Advanced Reservations. In: Proceedings of the IPDPS Conference, pp. 127–135 (May 2000)
22. Srivastava, J., Lurie, N.H.: A Consumer Perspective on Price-Matching Refund Policies: Effect on Price Perceptions and Search Behavior. Journal of Consumer Research 28(2), 296–307 (2001)
23. Vickrey, W.: Counterspeculation, Auctions, and Competitive Sealed Tenders. Journal of Finance 16, 8–37 (1961)
24. Wu, F., Zhang, L., Huberman, B.: Truth-telling Reservations. In: The 1st Workshop on Internet and Network Economics (2005)
25. Wurman, P.R., Walsh, W.E., Wellman, M.P: Flexible Double Auctions for Electornic Commerce: Theory and Implementation. Decision Support Systems 24, 17–27 (1998)
26. Yokoo, M.: The Characterization of Strategy/False-Name Proof Combinatorial Auction Protocols: Price-Oriented, Rationing-Free Protocol. In: Proceedings of IJCAI03, pp. 733–739 (2003)
27. Yokoo, M., Sakurai, Y., Matsubara, S.: Robust Double Auction Protocol against False-Name Bids. Journal of Desision Support Systems 39, 241–252 (2005)
28. Zheng, C.Z.: Optimal Auction in A Multidimensional World, Econometric Society World Congress 2000 Contributed Papers with number 0296 (2000)

Appendix A

This is appendix for calculating the optimal r^* in Fixed Cancellation Fee Auction (FCFA) mechanism. Let us consider the participation probability of a rational agent. Obviously, a rational agent will participate in the auction only when his expected utility $u_i = \theta_i(\beta_i - p) - (1 - \theta_i)r \geq 0$, then agent i's acceptable price p_i:

$$p_i \leq \beta_i - r(1 - \theta_i)/\theta_i \qquad (2)$$

Notice that $r \leq p_i$, and $\beta_i \leq 1$, substitute into above equation yields: $r \leq \theta_i \leq 1$ and $r/\theta_i \leq \beta_i \leq 1$. Therefore, for given cancellation fee r, the probability of a rational agent participated in the auction is

$$Pr(r) = \int_r^1 \int_{r/\theta}^1 d\beta d\theta = 1 - r + r\ln(r) \qquad (3)$$

Clearly, $Pr(1) = 0$ and $Pr(0) = 1$. We define $\psi(r, n)$ is the total expected number of n consumers willing to participate in the auction. Since the possibility of exactly k consumers participate in the auction can be described as:$Pr(r, k, n) = C_n^k Pr(r)^k (1 - Pr(r))^{n-k}$. The following equation is immediate:

$$\psi(r, n) = nPr(r) = \sum_{k=1}^n kPr(r, k, n) \qquad (4)$$

To calculate the expected revenue of the service provider obtain from these k consumers, we need predict the closing price p of the auction. Because we select m highest bids among these k requests, we should distinguish these two different situation: $k > m$ and $k \leq m$. Suppose there are exactly k potential consumers (whose $u_i \geq 0$) willing to submit their requests.

I) In the case of $k > m$, we sort prices $\{p_i\}$ into a decreasing order as $\{\hat{p}_i\}$, and $\hat{p}_{m+1}$ is the closing price. For any given price p and cancellation fee r, $r \leq p_i < p$ means that $r \leq \beta_i - r(1 - \theta_i)/\theta_i < p < 1$. So the probability $Pr(r \leq p_i < p)$ could be calculated as follows:

$$Pr(r \leq p_i < p) = \int_r^{\frac{r}{1-p+r}} \int_{r/\theta}^1 d\beta d\theta + \int_{\frac{r}{1-p+r}}^1 \int_{r/\theta}^{p+\frac{r(1-\theta)}{\theta}} d\beta d\theta \qquad (5)$$

The first integral item in the righthand of Equation (5) means $\theta < r\,(1 - p + r)$ when $p+r(1-\theta)/\theta > 1$, meanwhile the second integral item means $\theta \geq r\,(1-p+r)$ when $p + r(1 - \theta)/\theta \leq 1$. Similarly, the probability $Pr(p_i \geq p)$ is calculated as follows:

$$Pr(p_i \geq p) = \int_r^1 \int_{p+\frac{r(1-\theta)}{\theta}}^1 d\beta d\theta \qquad (6)$$

Now we describe the estimation of the mean value of $\hat{p}_{m+1}$. Suppose $\hat{p}_{m+1}$ is the (m+1)th highest price among k bidders, the distribution function the (m+1)th highest price $\hat{p}_{m+1}$ can be describes as:

$$F_{m+1}(p) = \sum_{t=0}^m C_k^t Pr(r \leq p_i < p)^t Pr(p_i \geq p)^{k-t} \qquad (7)$$

Let the density $f_{m+1}(p) = F'_{m+1}(p)$, then the expected mean value of (m+1)th highest value $\hat{p}_{m+1}$ is:

$$\tilde{p}_{m+1} = \int_r^1 p f_{m+1}(p) dp \qquad (8)$$

For example, the mean value of the second highest value among k bidders is $\int_r^1 pk(k - 1)(1 - F(p))F^{k-2}(p)f(r)dp$ [15]. Notice that if there are n random variables identically and independently uniformly distributed in $[0, 1]$, the mean value of i-th highest number's is $(n-i+1)/(n+1)$. For $p \in [r, 1]$, $\tilde{p}_{m+1}$ could be approximated using value $\tilde{p}_{m+1} = r + (k-m)(1-r)/(k+1)$. Then, the expected revenue of the seller is:

$$\pi_A(r, m, k) = \frac{m}{1 - r} \int_r^1 \theta \tilde{p}_{m+1} + (1 - \theta)r d\theta \qquad (9)$$

II) If $k <= m$, the expected revenue of the seller is:

$$\pi_B(r, m, k) = \frac{k}{1 - r} \int_r^1 \theta r + (1 - \theta)r d\theta = kr \qquad (10)$$

From I) and II), the expected revenue for given r could be calculated as follows:

$$ER(r, m, n) = \sum_{k=1}^m Pr(r, k, n)\pi_B(r, m, k) + \sum_{k=m+1}^n Pr(r, k, n)\pi_A(r, m, k) \qquad (11)$$

Therefore, the optimal cancellation fee r^* with respect to (n, m) is: $r^* = \arg max_r\{ER(r, m, n)\}$, and the maximal expected revenue is: $MER(m, n) = ER(r^*, m, n)$.

Appendix B

Let the valuations of consumer i as $\beta_{i,1}, \beta_{i,2}, \beta_{i,3}, \cdots$, where $\beta_{i,k}$ represents the marginal value of the k-th unit for i. More specifically, $\beta_{i,k}$ presents the increase of i's utility by obtaining one additional unit when i already has $k-1$ units, and for all i and k, $\beta_{i,k} \geq \beta_{i,k+1}$ holds, under the assumption that the marginal values decrease or remain the same. The assumption that marginal values decrease are widely adopted in economic models. We also assume that $p_{i,k} \geq p_{i,k+1}$ holds for all the bids of bidder i, where $p_{i,k}$ represents the bidding price of the k-th unit for i. A winner i who obtain k units pays the total of $\sum_{l=1}^{k} \max(p_{(l)}^{-i}, r)$, where $p_{(l)}^{-i}$ presents the l-th largest losing bid except those of i. This simple extension of the above to protocol could keep incentive compatible property when marginal values of all participates decrease or remain the same.

A Comparative Analysis of Single-Unit Vickrey Auctions and Commodity Markets for Realizing Grid Economies with Dynamic Pricing

Kurt Vanmechelen and Jan Broeckhove

University of Antwerp, BE-2020 Antwerp, Belgium
kurt.vanmechelen@ua.ac.be

Abstract. The introduction of market principles is a promising approach for dealing with the complex issues that arise in Grid resource management. A key aim is to align the resource consumption and provisioning patterns of Grid participants through proper incentive mechanisms. An important research question in this regard is the choice of a market organization. A number of such organizations have been proposed to support an economically inspired form of Grid resource management. This paper presents a comparative, quantitative, analysis of the single-unit Vickrey auctions and commodity market organizations with regards to price stability, fairness, and communicative and computational requirements. Our analysis based on simulated market scenarios shows that both market organizations lead to similar outcomes but that a commodity market organization leads to more stable market behavior at the cost of higher communicative requirements.

Keywords: Commodity Markets, Vickrey Auctions, Grid Economics, Resource Management, Grids.

1   Introduction

Traditional resource management systems adopt a system-centric form of resource management where a scheduling component establishes a mapping from jobs to Grid resources. This mapping is based on system oriented metrics such as infrastructure utilization or throughput. To generate broad support for Grids, but also to develop usage models that are more attuned to the user's needs, it is important that this emphasis shifts to a more user-centric approach. As such, the focus should be on allocation algorithms that are driven by the user's valuation of their results. In this way, Grids will deliver the maximum utility to the individual user. Because of their strategic and selfish nature, one cannot expect users to accurately formulate their true valuations to the resource management system unless proper incentive mechanisms are installed.

A promising approach towards dealing with this issue, involves the use of an economics based resource manager [1] which takes resource utilization cost

D.J. Veit and J. Altmann (Eds.): GECON 2007, LNCS 4685, pp. 98–111, 2007.

into consideration and requires users to back their valuations with associated credits, of which they have limited supply. Such an economics based trading model, where consumers rent resources from providers, is an attractive method to manage resource allocation in Grid systems. Aside from applications within the Grid domain [2,3,4,5,6,7,8,9,10,11], (consult [1] for an overview), economic models for resource sharing have also been applied to agent systems [12,13], telecommunication networks [14] and to databases [15] and data mining [16].

One of the most important research questions in adopting economic principles for Grid resource management is the choice of a market organization. Multiple such organizations exist in economic literature and at present, it is unclear which organization is most suitable to support an economically inspired form of Grid resource management. From a usage model point of view it is fairly clear that adopting *combinatorial auctions* [17], in which a participant can submit a single bid for a combination of goods, is one of the most attractive organizations. It enables consumers to accurately define their valuations for specific collections of Grid resources that are required by their applications. As such, it allows for expressing valuations that are conditional on the *coallocation* of a set of Grid resources. This eliminates the *exposure problem* [18] users face when they need to participate in multiple auctions for acquiring the constituent parts of an allocation bundle. However, this approach suffers from high computational complexity which can mostly be attributed to the NP-completeness of determining the optimal set of winners in such an auction [19]. In addition, the lower bounds on the communicative complexity of the value elicitation process in combinatorial auctions also inhibit their applicability for large scale economies, certainly in the case of general bidder valuations and when aiming for exact efficiency [20].

In this contribution we analyse the performance of two market organizations for realizing Grid economies; single-unit Vickrey auctions and commodity markets. The price formation process is fundamentally different in both organizations. In the commodity market one takes the approach of performing global optimization for establishing an equilibrium price, by polling all market participants for their supply and demand levels at a particular price. Participants are required to communicate these supply and demand levels to a central process performing the optimization, also called the *Walrasian Auctioneer*. The auction market organization, on the other hand, is fully decentralized and lets prices emerge from the local interactions of the market participants in single-unit Vickrey auctions. The goal of this contribution is to investigate whether these two approaches lead to different outcomes in terms of established prices, fairness of allocations and communicative and computational requirements for establishing these allocations.

Limited work has been done on directly comparing both systems on these grounds. The study in [9] compares both organizations on price stability and infrastructural utilization. The authors postulate that "auctioneering is attractive from an implementation point of view but that it does not produce stable pricing or market equilibrium, and that a commodity market performs better from the standpoint of a Grid as a whole". Similar remarks on the stability of prices in a single-unit Vickrey auction market are made in [5].

2 Market Model

For the purpose of this study we resort to simulation for efficiently analysing both market organizations on a large scale. Therefore, modeling decisions have to be made concerning the type of Grid resources that are simulated and the behavior of the market's participants. The model adopted here is similar to the one described in [11].

2.1 Resources

A complete and accurate Grid resource model should include a large set of different resource types, each with their own specific attributes. Examples include CPU time, scratch and permanent storage, network bandwidth, main memory, and more specialized resource types such as specific hardware components. Aside from taking a decision on the scope of the simulated resources, a second important design choice concerns the extent to and manner in which these resource types are introduced as tradeable goods into the market. Fully exposing all resource types and attributes allows for a very accurate valuation of resources by Grid users. A downside to this approach is the resulting increase in the complexity of the market's pricing mechanism, the interactions between the market and its participants, and the participants' valuation logic. In this contribution, we take the approach of restricting our resource model to CPU resources and to consider these the single type of resources that are tradeable in the resource market.

Commodity Market. In order to introduce diversification related to CPU performance, we introduce different commodity categories for multiple classes of CPUs with respect to their performance (in terms of GFlops/s). Each category is characterized by a *performance ratio* which expresses the performance increase of using a CPU from a particular category compared to a CPU from the lowest category. These categories constitute substitutable commodities, as jobs can execute on both, although they will be valued differently by consumers. The term *resource category* refers to a partition within a resource type based on a specific resource attribute, e.g. performance. Resources belonging to the same type but to a different category are substitutable. In this contribution we consider only one type and three categories.

Auction Market. In the auction market, all CPU resources will be individually auctioned. This allows for a more accurate valuation by consumers and is an advantage over the more abstract resource model used in the commodity market. Nevertheless, we will adopt the same single-attribute characterization from the commodity market in the form of the performance ratio, in order to keep results comparable.

2.2 Consumers

Each consumer has a queue of CPU-bound computational jobs that need to be executed and for which resources must be acquired from providers through participation in the market. The dispatch of a job to the CPU is effected immediately after the necessary resource has been acquired. Every job has a nominal running time T, i.e. the time it takes to finish the job on a reference CPU. However, in our spending algorithms we do not assume that the consumer has knowledge of this running time.

Every consumer is provided with a budgetary endowment that is periodically replenished. The period for this replenishment is denoted by an *allowance period*. We do not assume a particular funding source for the consumers. In practice, funding rates could be determined by system administrators, could be set by consumers themselves through monetary payment, or could result from a feedback loop that redistributes the credits earned by the providers of a particular (virtual) organization to the users of that organization. In every simulation step, consumers are charged with the usage rate prices for all Grid resources that are currently allocated to their jobs. Consumers do not attempt to save up credits, but try to use their entire budget. However, expenditures are spread evenly across the allowance period because we assume that consumers do not have reliable estimates for the running times of their jobs. Therefore, we need to prevent them from agreeing to a price, a "cost" level, that would not be sustainable for them over the entire allowance period.

Commodity Market. In the commodity market, consumers have to decide on the demand level they are willing to express, given a price vector P suggested by the market. The components of that vector P_i represent the price per resource unit, per time unit of the ith commodity category that is characterized by $PerformanceRatio_i$. Depending on the job mix a consumer has to schedule, certain resource categories will be preferred over others. This is expressed through the $ValuationFactor_i$ term. This leads to an adjusted price for each category, given by

$$AdjustedPrice_i = (P_i/PerformanceRatio_i)/ValuationFactor_i \qquad (1)$$

The r.h.s reflects the price normalized to unit performance and factors in the valuation. The consumer expresses demand, limited by the current allowed rate of expenditure, in the category with the lowest adjusted price.

The use of the $ValuationFactor_i$ term in the adjusted price is a simple abstraction for the complex logic a consumer might follow to prefer one CPU category over another. An example of such a logic whereby a consumer is willing to pay more than double the price for a CPU of category 1, which is only twice as fast as one of category 2, is the following. Suppose the consumer has a job graph that includes a critical path and that the user adopts a spending strategy for optimizing total turnaround time. Such a consumer would be willing to pay more than the nominal worth of a CPU of category 2 for allocating jobs on the critical path, as they have a potentially large effect on turnaround time.

Auction Market. For the auction market, consumers have to decide on the amount of credits they are prepared to bid for a particular CPU, the amount of CPUs they will bid on, and the specific auctions they will participate in. They calculate a base level for their bids which depends on the remaining budget and adjust this bid with the characteristics of the CPU resource:

$$Bid_{cpu} = (Base\_Bid * PerformanceRatio_{cpu}) * ValuationFactor_{cpu} \qquad (2)$$

The calculation for the base bid level also takes into account a target *parallelisation degree* the consumer wishes to realize. At the start of the simulation, all consumers try to launch all of their available jobs in parallel and determine their bids accordingly. As trading progresses, consumers gradually learn the level of parallellisation that they are able to achieve, given their budgetary limits, and they adjust their expectations and base level bids accordingly. Consumers adopt the simple heuristic of participating in the auctions which currently host the lowest number of bidders.

2.3   Providers

Every provider hosts a number of CPUs that can be supplied to the computational market. Once a resource is allocated to a job, it remains allocated until the job completes. The market price at the time of resource acquisition will be charged as a fixed rate to the consumer for the duration of the job. This approach is consistent with the fact that we do not assume a prior knowledge of a job's running time. An alternative to a fixed rate is to allow a variability in the charged rate based on the market price evolution. Another option is to allow variability on the performance a consumer receives for a given rate over the job's execution period, an approach adopted in [3]. These alternatives allow for potentially faster reallocation of resources according to the dynamic market environment, but make budgetary planning and resource usage planning more difficult for consumers.

For the analysis presented in this contribution, providers will not set minimum prices for their resources and will supply all of their available resources to the market.

2.4   Market Pricing

In the commodity market, prices for the different CPU categories are dynamically set in every simulation step by an optimizer which adjusts the price in order to bring the market to equilibrium. The optimizer iteratively polls all market participants for their supply and demand levels for each CPU category. This information is used to define an excess demand surface i.e. the difference between current demand and supply as a function of the price vector. An example of such an excess demand surface for a commodity market with two substitutable CPU categories is shown in figure 1. Note that we use the Euclidian norm of the excess demand vector.

The market equilibrium point is the zero of this surface and fixes the price at which the market will trade at that point in time. The global zero search algorithm is a combination of the algorithm presented in [11], which is an adaptation of Smale's algorithm [21], and a pattern search algorithm [22] of which we use the implementation provided by Matlab.

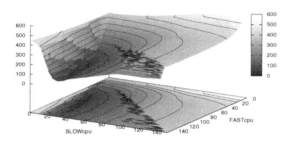

Fig. 1. Sample excess demand surface

In the auction market, each provider hosts a number of single-unit Vickrey auctions, one for each CPU that is available at that point in time. Consumers submit their sealed bids to the auctioneers of the CPUs they are interested in. The Vickrey auction allocates the CPU to the consumer with the highest bid, at the transaction price of the second highest bid (or zero if there is only one bidder). The fact that the consumer's transaction price does not depend on its own bid forms the basis for the *incentive compatibility* of the Vickrey auction. This means that a consumer has no incentive to place a bid which differs from its true value for the CPU, because no strategic advantage can be gained from this act.

3 Simulated Market Environment

We resort to a simulated market environment for analysing the commodity and auction market organizations. For this we use GES (Grid Economics Simulator) [23], a Java based discrete event simulator that we developed to support research into different market organizations for economic Grid resource management. The simulator supports both non-economic and economic forms of resource management and allows for efficient comparative analysis of different resource management systems. We currently have built-in support for commodity markets, different forms of auctions (English, Dutch, Vickrey, combinatorial and double auctions), fixed pricing as in [24], and implementations of other market mechanisms such as the proportional sharing approach found in Tycoon [3]. Non-economic resource management is supported through FIFO, round robin, and priority schedulers. The simulator is equipped with a user interface for supporting efficient analysis and configuration of market scenarios. A persistency framework allows for storing both scenario configurations and configurations of the UI layout. A screenshot of the UI is shown in figure 2.

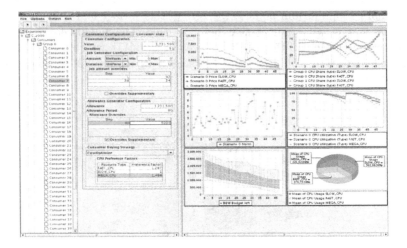

Fig. 2. Screenshot of the GES UI

The parameters of the scenario that we will use as the basis for our analysis are shown in table 1. For parameters that are specified with a range, we draw values from a uniform random distribution. Three groups of consumers with different budget levels are created by multiplying a consumer's base allowance with the respective *allowance factor* AF_i of its group. Note that in the context of this analysis, we keep the number of jobs in the consumer queues constant at the initial level by reinjecting a new job in the consumer's queue for every finished job. This results in a stable demand level which should lead to stable market prices.

Table 1. Simulation parameters

Parameter	Value
Number of consumers	100
Number of providers	50
Number of fastCPUs per provider	$\{1, 2, \cdots, 7\}$
Number of mediumCPUs per provider	$\{3, 4, \cdots, 11\}$
Number of slowCPUs per provider	$\{9, 10, \cdots, 17\}$
Performance ratio of fast vs slow	3.0
Performance ratio of medium vs slow	2.0
Valuation factors	[1.0,1.5]
Job running time in time steps	$\{4, 5, \cdots, 8\}$
Number of jobs per consumer (constant)	$\{150, 151, \cdots, 500\}$
Base Allowance	100,0000 * [1.0,1.5]
$\{AF_1, AF_2, AF_3\}$	$\{1.0, 2.0, 3.0\}$
Allowance period in time steps	800

4   Comparative Analysis

4.1   Dynamic Pricing

As shown in figure 3, the average prices paid by the consumers in the market for the different categories of resources are similar. The auction market shows a higher fluctuation in the price levels over the course of the simulation with a relative standard deviation [25] of 5.86% versus 1.62% for the commodity market. The deviation for the auction market prices does not include the instable price levels of the first 10 steps, if these are included the deviation increases to 9.62%. Whereas the commodity market immediately brings the market to equilibrium through global optimization, the participants in the auctions still have to optimize their *target parallellisation degree* and discover the amount of resources to bid for. This results in the extensive adjustments of the average CPU price paid at the beginning of the simulation.

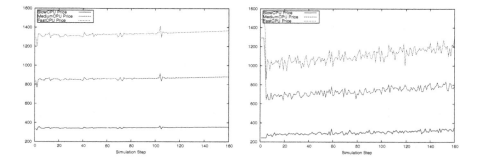

Fig. 3. CPU Prices for the commodity market (left) and auction market (right)

Although prices are slightly less stable in the auction market, they do follow the trend of supply and demand in the market as shown in figure 4. This scenario is similar to the one used in [9]. Periods of overdemand are followed by periods of underdemand through the injection of a set of jobs into the system at intervals of 45 simulation steps. In addition, jobs are not reinjected in the consumer queues on completion. Whereas the results in [9] indicate that such a scenario leads to very erratic pricing behavior for the auction market, showing price levels that do not reflect overall market supply and demand, the results are very different here. Although some parameters of the simulation differ, one of those being the fact that consumers have to coallocate disk and CPU resources in [9], our results do show that it is possible, using fairly simple bidding logic, to obtain meaningful and fairly stable average prices in the auction market. We note that the two price peaks for the slow CPU category in the commodity market scenario are caused by the fact that no slow CPU resources are available for trade at those time instances. The equilibrium optimizer generates a high price level for these

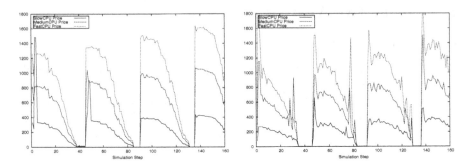

Fig. 4. CPU Prices in the varying load scenario for the commodity market (left) and auction market (right)

resources in order to remove all demand for them in the market and minimize excess demand.

We note that the average transaction price for resources is lower in the auction market. The difference in total revenue generated for the providers is 15.79%.

4.2   Fairness

The fairness of the allocations in an economic resource management system, denotes whether the level of budgetary endowment of a consumer is correctly translated into a corresponding share of the infrastructure allocated to that consumer. The graphs in figure 5 show that in both markets the average infrastructural shares of the three consumer groups converge to the budget shares of those groups. In the auction market, correspondence is not achieved in the first simulation steps. This can be explained by the fact that consumers are still learning the parallellisation degree that is sustainable by them in the current market situation. The commodity market does not require such judgment from its consumers and immediately brings about fair allocations. To investigate whether shares quickly adapt to sudden changes in the market, we swap the budget levels of the different consumer groups at step 80. As shown in figure 6, both market organizations are able to quickly adapt the allocations to reflect the new market situation. Instead of the cumulative average, figure 6 shows the instantanious shares at each simulation step, which are somewhat less stable for the auction market scenario.

From the providers' point of view, a fair operation of the market should lead to similar revenues among the different providers for the CPU resources sold. In the commodity market, we measured an average relative standard deviation of 2.09% on the nominal transaction price paid per CPU cycle in a single time step. This price is obtained by dividing the earned revenue on a set of CPUs by the performance factors of these CPUs. The origin of this deviation lies in the valuation factors consumers have towards different CPU categories and the differences in the number of CPUs each provider has of a particular category.

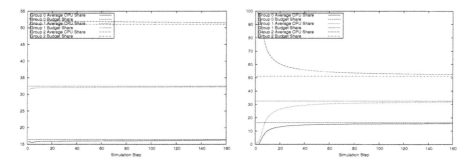

Fig. 5. Budget and infrastructural shares as a cumulative average for the commodity market (left) and auction market (right)

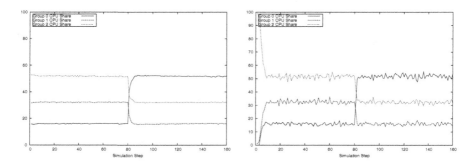

Fig. 6. Fairness under sudden budget change for the commodity market (left) and auction market (right)

In the auction market, prices emerge from the local interactions among the bidders and this results in less stable revenue streams for the providers. The average deviation was 6.67% in the auction market. Although a greater variance in revenue was observed on the transactions made at a single step of the simulation, the deviation on the total nominal revenue accrued by the providers at the end of the simulation was only 2.33%. For the commodity market this relative standard deviation was 2.06%.

In the less stable peaked demand scenario, the differences in revenue stability between the two market organizations increase. For the auction market, we measured a deviation of 43.74% on transaction prices for a single step and a deviation of 6.79% on the total accrued revenue. The respective deviations for the commodity market under the peak demand scenario were 4.03% and 2.31%.

4.3 Computational and Communicative Requirements

The number of resource categories introduced in the commodity market is a determining factor for its communication and computational requirements. This can be attributed to the fact that each additional CPU category increases the dimensionality of the equilibrium price optimization problem. Tables 2 and 3 show the effect of introducing more categories for the constant load and peaked load scenarios. The number of messages and running time are given per simulated time step, as well as the median of the excess demand norm over all simulation steps. The tables also include the number of network messages sent in the auction market for comparison.

Table 2. Market performance with respect to the number of CPU categories in the market (constant load scenario)

CPU categories	CM Messages	Running Time (ms)	Median norm	Auction Messages
1	4440	80	1.0	2996
2	6276	314	1.41	2541
4	47033	1187	5.48	1962
6	90812	2355	6.86	1675
8	157608	4539	8.94	1625

Table 3. Market performance with respect to the number of CPU categories in the market (peaked load scenario)

CPU categories	CM Messages	Running Time (ms)	Median norm	Auction Messages
1	3482	50	1.0	1646
2	33257	314	3.16	1576
4	93177	1084	7.55	1359
6	163750	2629	12.79	1312
8	261181	4808	72.02	1250

Introducing more categories allows the market participants to express their valuations for the different levels of CPU performance in a more fine grained manner. As a result, resource allocations will be better adapted to the real needs of the users. Although the computational costs do not inhibit a practical deployment of this market organization (they would allow for adapting the price within a timeframe of 5 seconds for the case of 8 CPU categories), the communicative requirements of the price optimization process might. This is especially true for large scale, wide-area deployments with higher communication delays, lower network bandwidth and higher network usage costs. Nevertheless, the communication burden can be reduced significantly through a dynamic deployment of the consumer bidding logic into the local environment of the price optimization process. In a Java based environment, this can be realized by allowing market participants to send an object representing their *bidding agent* to the JVM of the equilibrium optimizer when new prices are to be formed. The agent then reports the participant's supply or demand levels to a local component which aggregates these levels and passes them on to the optimizer through local method calls. Java's support for dynamic classloading allows the agent's code to be automatically downloaded when needed. This model has already been validated through a real-world deployment of the commodity market logic using the CoBRA framework [26]. Organizing multiple aggregators in a tree structure can further address the scalability issues of the commodity market. We also note that the median of the norm, which is a measure for the excess demand after commodity market prices are set [10], increases as we introduce more categories.

The amount of network messages sent in the auction market organization is significantly lower, especially for the scenarios with a higher number of categories. It diminishes as we introduce more categories because we keep the total processing capacity of the Grid constant, while introducing more types with higher performance factors. This leads to a lower amount of discrete resources that need to be auctioned. Note that an auction based framework which uses English auctions for example, can lead to significantly higher communication costs as a result of iterative overbidding in such auctions. Because of their strategy-proofness, single-unit Vickrey auctions require only one round of bidding, resulting in the minimum amount of communication necessary to establish a trade. Another important factor for the lower amount of communication is the fact that a consumer only participates in a limited number of auctions (according to its target parallellisation degree). On average, each auction attracted approximately five consumers in the simulated scenarios.

5   Conclusion

Both Vickrey auctions and commodity markets have been proposed as market organizations for establishing Grid resource management systems that are based on economic principles. In order to guide system designers in their choice for a particular organization, we have presented a comparative analysis of both options on the grounds of price stability, fairness and communicative and computational requirements. The commodity market organization results in a more stable environment with respect to prices and allocative shares. The main disadvantages of this organization are its limited support for fine-grained valuations because of the high communication costs when defining a large number of resource categories, and its centralized nature. The Vickrey auction organization leads to similar but less stable outcomes and supports fine-grained valuations at significantly lower communicative requirements.

References

1. Buyya, R., Abranson, D., Venugopal, S.: The Grid Economy. In: Proceedings of the IEEE, vol. 93(3), pp. 698–714 (2005)
2. Abramson, D., Buyya, R., Giddy, J.: A computational economy for Grid computing and its implementation in the nimrod-g resource broker. Future Generation Computer Systems 18(8), 1061–1074 (2002)
3. Feldman, M., Lai, K., Zhang, L.: A price-anticipating resource allocation mechanism for distributed shared clusters. In: EC '05: Proceedings of the 6th ACM conference on Electronic commerce, pp. 127–136 (2005)

4. AuYoung, A., Chun, B.N., Snoeren, A.C., Vahdat, A.: Resource allocation in federated distributed computing infrastructures. In: Proceedings of the 1st Workshop on Operating System and Architectural Support for the On-demand IT InfraStructure (2004)
5. Waldspurger, C.A., Hogg, T., Huberman, B.A., Kephart, J.O., Stornetta, W.S.: Spawn: A distributed computational economy. IEEE Trans. Softw. Eng. 18(2), 103–117 (1992)
6. Gomoluch, J., Schroeder, M.: Market-based resource allocation for Grid computing: A model and simulation. In: Int'l. Middleware Conference, Workshop Proceedings, pp. 211–218 (2003)
7. Regev, O., Nisan, N.: The POPCORN Market – an Online Market for Computational Resources. In: Proceedings of the 1st International Conference on Information and Computation Economies, pp. 148–157 (1998)
8. Chun, B.N., Culler, D.E.: User-centric performance analysis of market-based cluster batch schedulers. In: Proceedings of the 2nd IEEE/ACM Int'l. Symposium on Cluster Computing and the Grid, pp. 22–30 (2002)
9. Wolski, R., Brevik, J., Plank, J.S., Bryan, T.: Grid Resource Allocation and Control Using Computational Economics. In: Grid Computing: Making the Global Infrastructure a Reality, pp. 747–772. Wiley and Sons, Chichester (2003)
10. Vanmechelen, K., Stuer, G., Broeckhove, J.: Pricing Substitutable Grid Resources using Commodity Market Models. In: Proceedings of the 3th International Workshop on Grid Economics and Business Models, pp. 103–112. World Scientific, Singapore (2006)
11. Stuer, G., Vanmechelen, K., Broeckhove, J.: A Commodity Market Algorithm for Pricing Substitutable Grid Resources. Fut. Gen. Comp. Sys. 23(5), 688–701 (2007)
12. Bredin, J., Maheswaran, R.T., Çagri Imer, T., Başar, D.K, Rus, D.: Computational markets to regulate mobile-agent systems. Autonomous Agents and Multi-Agent Systems 6(3), 235–263 (2003)
13. Dash, R., Parkes, D., Jennings, N.: Computational Mechanism Design: A Call to Arms. IEEE Intelligent Systems 18(6), 40–47 (2003)
14. Gibney, M.A., Jennings, N.R., Vriend, N.J., Griffiths, J.-M.: Market-based call routing in telecommunications networks using adaptive pricing and real bidding. In: Proceedings of the 3th Int'l. Workshop on Intelligent Agents for Telecommunication Applications, pp. 46–61 (1999)
15. Stonebraker, M., Aoki, P.M., Litwin, W., Pfeffer, A., Sah, A., Sidell, J.: Mariposa: A wide-area distributed database system. VLDB J. 5(1), 48–63 (1996)
16. Joita, L., Rana, O., Freitag, F., Chao, I., Chacin, P., Navarro, L., Ardaiz, O.: A catallactic market for data mining services. Future Generation Computer Systems 23(1), 146–153 (2007)
17. Cramton, P., Shoham, Y., Steinberg, R. (eds.): Combinatorial Auctions. MIT Press, Cambridge (2006)
18. Bykowsky, M., Cull, R., Ledyard, J.O.: Mutually Destructive Bidding: The FCC Auction Design Problem. Journal of Regulatory Economics 17(3), 205–228 (1995)
19. Lehmann, D., Mller, R., Sandholm, T.: The Winner Determination Problem. In: Cramton, P., Shoham, Y., Steinberg, R. (eds.) Combinatorial Auctions, pp. 297–317. MIT Press, Cambridge (2006)
20. Lehmann, D., Mller, R., Sandholm, T.: The Communication Requirements of Combinatorial Allocation Problems. In: Cramton, P., Shoham, Y., Steinberg, R. (eds.) Combinatorial Auctions, pp. 265–294. MIT Press, Cambridge (2006)
21. Smale, S.: A Convergent process of price adjustment and global newton methods. Journal of Mathematical Economics 3(2), 107–120 (1976)

22. Lewis, R., Torczon, V.: A globally convergent augmented Lagrangian pattern search algorithm for optimization with general constraints and simple bounds. SIAM Journal on Optimization 12(4), 1075–1089 (2002)
23. Vanmechelen, K., Stuer, G., Broeckhove, J.: Grid Economics Simulator (GES), CoMP Group, University of Antwerp, available from the authors on request (2007)
24. Buyya, R.: Economic-based Distributed Resource Management and Scheduling for Grid Computing, Ph.D. dissertation, Monash University, Australia (2002)
25. Snedecor, G., Cochran, W.G.: Statistical Methods, 6th edn. pp. 62–64. The Iowa State University Press, Ames, IA (1967)
26. Hellinckx, P., Stuer, G., Hendrickx, W., Arickx, F., Broeckhove, J.: User Driven Grid Research, the CoBRA Grid. In: Proceedings of the International Workshop on Grid Testbeds, colocated with CCGrid06, Singapore, pp. 1–8 (May 16-19, 2006)

A Continuation-Based Framework for Economy-Driven Grid Service Provision

Maurizio Giordano and Claudia Di Napoli

Istituto di Cibernetica "E. Caianiello", C.N.R.
Via Campi Flegrei 34, 80078 Pozzuoli, Naples, Italy
{m.giordano,c.dinapoli}@cib.na.cnr.it

Abstract. The management of computational resources is a crucial aspect in grid computing because of the decentralized, heterogeneous and autonomous nature of these resources that usually belong to different administrative domains and are provided in dynamic and changing environments. For this reason more sophisticated computing methodologies are necessary to provide these resources in a flexible manner. In particular, the possibility of controlling the execution of services in grid is a crucial aspect in order to change service execution policies at run-time.

In the present work an infrastructure to model service providers is proposed to allow for flexible provision of grid services, i.e. to allow providers to dynamically control the execution of services according to the changing conditions of the environment where they operate in. The infrastructure is based on *continuations*, a programming paradigm that allows to control the state of program execution at application-level without involving the operating system stack. This approach makes the proposed infrastructure a flexible and easily programmable middleware to experiment different scheduling policies in service-oriented scenarios.

Keywords: Grid service provision, continuations, quality of service.

1   Introduction

Computational grids represent the new research challenge in the area of distributed computing. They aim at providing a unified computational infrastructure composed of networked heterogeneous resources that makes effective use of the computational power delivered by each resource.

A computational grid is a pool of resources that are not subject to centralized control (i.e. that live within different control domains and that do not rely on a central management system), that use standard, open, general-purpose protocols and interfaces (i.e. not application-specific). Resources can be combined in order to deliver added value services so that the utility of the resulting system is significantly greater than that of the sum of its parts. Users will be able to access and share these computing resources on demand over the Internet, relying on an infrastructure that is expected to be resilient, self-managing, and always available, and above all that is perceived as a unified framework by end

D.J. Veit and J. Altmann (Eds.): GECON 2007, LNCS 4685, pp. 112–123, 2007.

users. In order to provide such a computational infrastructure, grid technologies should support the sharing and coordinated use of diverse resources in a dynamic environment [1].

In addition, in grid environments resource providers (that can be individuals, organizations, groups, government, and so on) are independent and autonomous entities that need to be motivated to make available the resources they provide.

A market-oriented approach can be used to provide the possibility of buying and selling computational resources in the same way as goods and services are bought and sold in the real world economy [2]. Adopting a computational economy-based view [3,4] where resources are provided at a given cost constitutes *per se* a mechanism for encouraging resource owners to contribute their resources for the construction of the grid, and compensate them based on the resource usage, i.e. on the value of the work done. So, the ultimate success of computational grids as a production-oriented commercial platform for solving problems is critically dependent on the support of market/economy-based mechanisms to resource management.

In such production-oriented (commercial) computational grids, resource owners act as service providers that make a profit by selling their services to users that act as buyers of computational resources for solving their problems.

In this scenario, service providers need to have control on the execution of the services they provide in order to accommodate for the changing *Quality-of-Service* (QoS) [5] requirements service consumers can ask for.

In this work we propose an infrastructure to model service providers to control the execution of services by allowing for service suspension and resuming in a way similar to process preemption and control in traditional operating systems.

The infrastructure relies on *continuation* programming paradigm [6] in order to provide execution state saving and restoring mechanisms for services. These mechanisms will support the possibility of explicitly controlling the execution of services to allow providers to decide "how" to fulfil a service request, i.e. what Quality-of-Service to provide at run-time.

The rest of the paper is so organized: in section 2 the economic-based service-oriented scenario is described as the reference application domain; section 3 reports the proposed service provider architecture together with its functionalities and interfaces; in section 4 some use-cases are outlined to show the applicability of the proposed infrastructure in commercial computational grids; finally section 5 reports some conclusions.

2   A Service-Oriented Approach for Economic-Aware Grids

In the present work a service-oriented approach is adopted as described in [1], where grid resources are exposed to the network as *grid services*, i.e. computational capabilities defined through a set of well-defined interfaces, and a set of standard protocols used to invoke the services from those interfaces, and it has to be identified, published, allocated, and scheduled. A service is provided by the *body* responsible for offering it, we refer to as *service provider*, for consumption by others, we refer to as *service consumers*, under particular conditions. In this view, a service provider represents the interface between a service consumer and a required functionality, i.e. a grid service.

It is worth noting that in our scenario we refer to a grid service as any type of computational resource made available through the network according to platform independent interfaces and protocols, so it can also be a web service compliant with OGSA [7]. In the rest of the paper we refer to "web services", or "grid services", or simply "services" assuming they have, from our point of view, the same meaning.

It is well recognized that in a market-based service-oriented grid, services will be provided with some user-dependent *Quality-of-Service* (QoS) requirements, i.e. with characteristics that meet expectations and obligations agreed between the provider and the consumer. Service providers may want to optimize utilization, i.e. their profit, whereas service consumers may want to optimize response time while minimizing cost. So, the same service could be provided with different QoS.

It is beyond the scope of the present work to study how complex the quality of a service can be, and how to characterize it, i.e. how many parameters should be considered to express the quality of a service, and how it can be represented, that is mainly a domain-specific problem. In general, we assume that a service request is fulfilled when the user requirements on the Quality-of-Service can be met by the service provider that received the request. According to the current research directions, the match is stated in Service Level Agreements (SLAs) [8], that represent bilateral agreements typically between a service provider and a service consumer established before service execution.

Nevertheless, it is likely that in very dynamic and changing computing environments like the grid, service providers can make different decisions on the Quality-of-Service they provide their services with according to the requirements of new service requests, e.g. they may want to break or change some agreements in the case a new consumer comes with a more remunerative request. Also service consumers may decide to change some requirements on service execution, e.g. they may want to pay more to have a service delivered earlier. In such cases, it is advisable to control the execution of services on demand by suspending and resuming their execution according to decisions made at run-time.

3   A Continuation-Based Service Provider

In order to be able to control the execution of services, we propose a service provider architecture supporting web service *preemption* facilities for the suspension and resuming of service instance execution based on *continuation* management [6].

The main feature of this architecture is the possibility to use a set of primitives to control service execution, i.e. to submit, suspend, resume web services. The primitives allow to specify QoS parameters affecting the service scheduling decisions (priorities, cpu resource access, and so on) and to change them at run-time. These primitives are also exposed as web services that can be invoked by any client program acting on behalf of service brokers and/or service consumers and they are accessed in a distributed setting through XML/SOAP messaging.

In this way we provide a uniform mechanism to control service execution policies at two levels: a *low level* where service providers control their own service execution according to local policies, and a *meta-level* where global decisions need to be made for the coordination of services provided by different providers. In the latter case, we foresee that the primitives to control service execution can be used by a *metascheduler* [9,10] in charge of coordinating the local schedulers of different providers.

The primitives are based on the possibility to capture the state of a computation by means of continuations, so the computation can be suspended and resumed later on.

3.1   The Notion of Continuation

A *continuation* relative to a point in a program represents the *remainder of the computation* from that point [6], so a continuation is a representation of the program current execution state. Continuation capturing allows to package the whole state of a computation up to a given point. Continuation invocation allows to restore that previous state restarting the computation from that point. Although any programming system maintains the current continuation of each program instruction it evaluates, these continuations are generally not accessible to the programmer.

In functional programming languages, the continuation can be represented as a function and the possibility of explicitly managing it allows to effect the program control flow [11]. In languages like C the current execution state is represented by the call stack state, the globals, and the program counter. Some object-oriented programming languages support continuations by providing constructs to save the current execution state into an object, and then to restore the state from this object at a later time.

In our implementation we used Stackless Python [12], an experimental implementation of the Python programming language that uses continuation support to model concurrency in an easy way. It provides abstractions of microthreads at application level, named *tasklets*, whose implementation is based on continuations. Stackless Python supports *tasklets* as built-in user-level lightweight threads with constructs to control their creation, suspension, resuming and scheduling at application level. Furthermore, Python is one of the languages that provides a satisfactory support of libraries and tools for the development of web services [13].

3.2   The Service Provider Architecture

In order to be able to provide services that meet Quality-of-Service requirements both of service consumers (e.g. cost, response-time) and of providers (e.g.

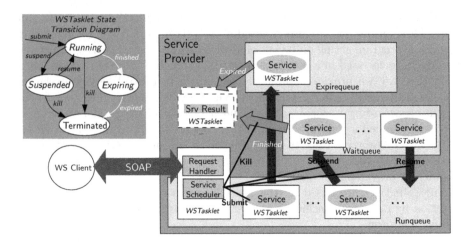

Fig. 1. Services provider architecture and service state transition

throughput, profit, CPU utilization), it is crucial to be able to control the execution of services in accordance with new events occurring in the environment since these requirements cannot be statically determined.

Service preemption mechanisms are a way to provide full control of service execution and they can be implemented (or simulated) using several approaches, both at application or operating system level. For examples, at application level the Java language provides (deprecated) thread suspension/resuming support. Other approaches [14] use operating systems signals (SIGSTOP/SIGCONT) available in most operating system infrastructures.

The main objective of the proposed service provider architecture is application-level preemption of services in order to support at programming level the development of dynamic policies for service execution. Service preemption is not provided at operating system level, but at application-level by managing program continuations. This choice makes the framework flexible and easily adaptable for developing and experimenting scheduling facilities, policies and service-control in different service-oriented architecture applications.

Existing web service frameworks [15,16] make it difficult to implement a service provider architecture with preemption mechanisms of web services without a deep changing of the control patterns usually implemented as a built-in feature. This is because they usually obey to the *Inversion of Control* (IoC) programming pattern [17,18] widely used in most Java and object-oriented web application environments. So, web service instantiation and life-cycle management cannot be fully controlled by programmers who develop and add web services to the framework.

For this reason existing web service frameworks are not suitable to provide an application-level control of service execution supporting service suspension and resuming.

For this reason we designed a service provider equipped with mechanisms to process suspension and resuming notifications. The service provider should process, from time to time, arrival of notification messages in order to suspend/resume the execution of a service it is providing by capturing/restoring its continuation. The control of service execution can be driven both by the service provider itself and by any client program. Service preemption, driven or not by client requests, is carried out by the provider storing at the preemption points the execution state of the specified service.

The client program can represent either a service consumer that requires a service result, or a metascheduler or a service broker trying to adapt local service execution policies so that resources can be shared in a reliable and efficient way in a heterogeneous and dynamically changing environment like the grid.

The service provider architecture is depicted in figure 1 and it is implemented in Stackless Python. The provider is represented by a *service container* consisting of a pool of lightweight threads, named *WSTasklets*, implemented by using continuations. WSTasklets execute concurrently in the same Python interpreter process.

WSTasklets are threads wrapping up service functions that represent web service *operations* in WSDL [19]. Web service operations are given as parameters to a WSTasklet wrapper and executed within its context (see figure 1). The wrapping guarantees the required functionalities to suspend/resume web service operation executions by means of the Stackless Python continuation storing and resuming support.

A WSTasklet, and hence the corresponding service, can be in the following states:

- *running*, i.e. the WSTasklet is in execution or ready to be scheduled for execution;
- *suspended*, i.e. the WSTasklet is not yet terminated, but cannot be scheduled for execution;
- *expiring*, i.e. the WSTasklet terminated its execution, but its descriptor is still alive to make the result available to successive requests;
- *terminated*, i.e. the WSTasklet terminated its execution and its descriptor is no longer available because either a specified expiration time elapsed, or the client requested and obtained the result before the expiration time.

There is a special WSTasklet, always in *running* state, that is the main thread of the service provider. It interleaves messaging and scheduling activities by means of two modules: the *Request Handler* and the *Service Scheduler*. The *Request Handler* deals with probing incoming SOAP messages; the *Service Scheduler* module controls WSTasklet state transitions by means of a set of primitives: *submit, suspend, resume, kill* (black arrows in the diagram shown in figure 1). The *submit* primitive creates a new WSTasklet, wrapping up a specified service operation and puts it in the *running* state.

The Service Scheduler maintains three queues to manage WSTasklets in different states:

Runqueue - it contains all WSTasklets running or ready to be scheduled for execution. Threads in this queue are by default executed in time-sharing mode by assigning to each WSTasklet a *time quantum* that can be changed by the Service Scheduler (also in response to incoming SOAP requests).

Waitqueue - it contains all WSTasklets suspended and thus removed from the Runqueue. The provider may decide to suspend/resume service execution according to both its own scheduling policy, and upon receiving specific SOAP requests from an external application, e.g. a metascheduler.

Expirequeue - it contains all WSTasklets that finished executing and that are waiting to be garbage-collected by the system. They are maintained in this queue in order to keep the computation results that can be later on collected by service consumers with SOAP requests within a given *expiration time*. The expiration time is not necessarily a system specific parameter, and it could be specified as a QoS parameter at the submitting phase.

It should be noted that in the Service Scheduler module different scheduling policies can be implemented at application-level overriding the default one both by changing the *time quantum* and by re-organizing the Runqueue. In this way the service provider is able to change its own local scheduling policy at run-time directly invoking the primitives to control service execution.

3.3   Asynchrnonous Client-Provider Interaction

As outlined earlier, the proposed infrastructure allows also to access the primitives to control service execution as web services to be invoked by any external client program. In such a case, a client-provider interaction takes place and it is implemented as an asynchronous request/response operation with polling [20]. Asynchronicity allows the client to proceed the computation concurrently with the web service execution until the operation result is required: at this point the client needs to synchronize with the provider and establishes a new communication to retrieve the result.

We extend the asynchronous request/reply operation mode with functionalities to suspend and resume web service execution. The client-provider asynchronous interaction pattern is described in figure 2 where a client invokes a web service operation, named "Operation A", offered by the continuation-based service provider.

The primitives to control service execution are exposed as the following WSDL operations: `submit`, `suspend`, `resume` and `probe`. They represent *meta-operations* because they are invoked by clients to control and to monitor web service operation executions.

The client-provider interaction pattern is started by clients invoking the `submit` WSDL operation to request a service execution. The `submit` request invokes the "Operation A" on a set of input arguments and starts its execution (see the syntax in figure 3(a)). The provider sends back to the client a reply with an acknowledge that the submission is done together with a *correlation ID*. The correlation ID is unique and is set by the provider to be used together with

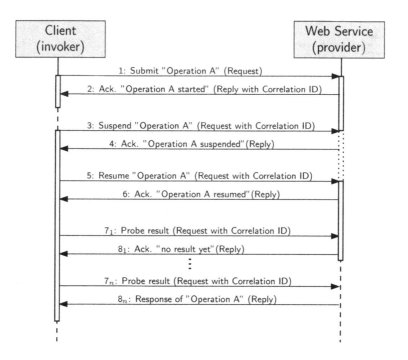

Fig. 2. Asynchronous request/response operation with polling and suspend/resume facility

the client to associate subsequent requests and responses belonging to the same client-provider transaction.

Correlation tokens to embed multiple messages in transactions are widely used in most asynchronous web service protocol proposals [21,22]. Approaches differ for the particular protocol adopted (JMS, SOAP, etc.) and/or the mechanisms used to implement message correlation. In our approach correlation is explicitly included in SOAP message bodies as shown in figure 3.

The **submit** request includes a set of **qos** parameters. QoS attributes are specified by clients to drive or affect scheduling policy of the web service operation execution.

The client starts the execution of "Operation A" and continues its computation so that it may also decide to suspend the web service execution, to resume it later on up to completion.

To perform suspend and resuming actions the client uses the **suspend** and **resume** meta-operations. The **suspend** request uses the correlation ID to refer to the web service operation (instance) to be suspended. Upon receiving the request, the provider captures and saves the execution state of "Operation A", and it sends back to the client an acknowledge.

The **resume** request uses the correlation ID to refer to the web service operation (instance) whose execution must be resumed. Upon receiving the request, the provider retrieves the execution state (continuation) stored and tagged with

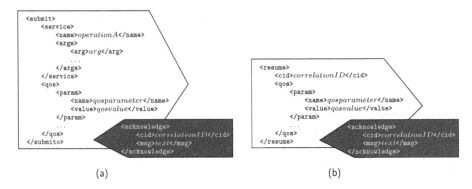

```
<submit>
    <service>
        <name>operation A</name>
        <args>
            <arg>arg</arg>
            ...
        </args>
    </service>
    <qos>
        <param>
            <name>qosparameter</name>
            <value>qosvalue</value>
        </param>
        ...
    </qos>
</submito>
```
```
<acknowledge>
    <cid>correlationID</cid>
    <msg>text</msg>
</acknowledge>
```

```
<resume>
    <cid>correlationID</cid>
    <qos>
        <param>
            <name>qosparameter</name>
            <value>qosvalue</value>
        </param>
        ...
    </qos>
</resume>
```
```
<acknowledge>
    <cid>correlationID</cid>
    <msg>text</msg>
</acknowledge>
```

(a)   (b)

Fig. 3. Service control primivites syntax: (a) `submit`; (b) `resume`

the specified correlation ID. It then resumes the web service operation and sends
back to the client an acknowledge.

As described in figure 3(b), also the `resume` request includes specifications of
QoS parameters. This means that in our framework a service execution can be
resumed by changing at run-time the web service operation scheduling policies.

Client-provider synchronization is implemented by the `probe` request. The
`probe` checks if "Operation A" is finished. If the request occurs before the web
service operation exits (the first `probe` of figure 2), the client is acknowledged
that the result in not ready yet. After a successful `probe` request the client
synchronizes with the provider and gets the result.

4   Application Scenarios

Economy-based grid environments require more sophisticated scheduling approaches able to deal with several optimisation functions: those provided by the
user with his/her objectives (e.g. cost, response-time) as well as those objectives
defined by the resource providers (e.g. throughput, profit, CPU utilization). It is
also important to be able to change these objectives according to new conditions
occurring when fulfilling service requests.

Our framework provides this flexibility since it is possible to associate to the
request of a service execution a `qos` parameter taking into account the cost of
a service and to allow both the client and the provider to use its value to drive
service execution suspension and resuming.

For example, let's suppose that a client requests the execution of a service
with an associated *cost* representing an estimate of the amount of resource the
client expects the web service will consume (e.g. CPU time). Thus the *cost*
may correspond to the maximum execution time guaranteed by the provider,
according to a previous agreement with the consumer. In this case the service
request can be `submitted` with a `qos` parameter value corresponding to the *cost*.
The Request Handler module of the service provider receives the request with
the associated cost and the Service Scheduler starts its execution in time-sharing.

While the time-sharing quantum is fixed for all services, the amount of quanta available for the service is limited by the *cost* parameter. If the service has not completed before this time, the Service Scheduler suspends it.

Depending on the client-provider agreement, the suspended service can be rescheduled for execution only after all tasks have finished spending the time slices they paid for, or it can be resumed only if the client pays an additional cost. In the latter case the client `resumes` the operation with a new value of the `qos` parameter guaranteeing an additional execution time slice for it. In this scenario the Service Scheduler makes scheduling decisions according to the received requests and it controls the execution of the required services accordingly.

The possibility of changing the scheduling policies at run-time can be exploited also when the cost of service execution is dependent on the priority the provider assigns to the service it provides. We assume that when consumers request a service execution, the provider charges for the service a cost according to the priority which the service will be executed with. In this scenario it is possible for consumers to increase/decrease, at any time, the money they are willing to pay for the the required services. In such a case the provider should respectively increase/decrease the priority it assigned to the service execution requested by that consumer. In this scenario the `qos` parameter included in service submission represents the priority assigned to it; if the consumer wants to change it, it request the provider first to suspend the service and then to resume it with a changed priority value.

Of course, in both scenarios it is up to the local Service Scheduler, according to the adopted policy, to account for the cost/priority change request and to fit it in the multitasking environment.

5   Conclusions

In this work we propose a service provider architecture based on continuations storing and management to provide primitives to control web services execution and to implement service scheduling policies. The primitives are also offered by the service provider to external client applications via web service (SOAP/WSDL) interfaces.

With this approach we may implement the service execution policies at two levels: the lower level relies at the service provider layer to implement local schedulers; the higher level can be a metascheduler that interacts with multiple service provider schedulers in a distributed setting by means of SOAP messaging.

Community Scheduler Framework (CSF) [23] is an infrastructure providing facilities to define, configure and manage *metaschedulers* for the grid. Metascheduling is conceived as a higher level of scheduling decisions to coordinate local schedulers (PBS [24], LSF [25]) on hosts and clusters in a grid environment.

Our approach has similarities to CSF. Both solutions pursue the scope of providing new high-level scheduling functionalities both to service consumers and to the development of metascheduler middleware. CSF functionalities mainly target configuration and management of scheduling policies and their coordination in a grid environment.

CSF provides scheduling functions both via web service (SOAP/WSDL) interfaces and by means of client interfaces or shell commands. Like CSF, our framework allows service execution control and scheduling queues configuration and management through SOAP-based messaging interaction.

Although both approaches support suspension/resuming facilities, CSF applies them to control jobs, i.e. processes running in the hosting OS environments. CSF defines high-level scheduling services (in Java) to drive and translate consumer requests into job-control commands implemented at lower level in the scheduler running on the target hosts or clusters (as PBS and LSF). Therefore CSF job-control functionality depends on the underlying OS layer service compatibility.

In the present work we developed a continuation-based service provider featuring programmable and full control of generic web service executions. The service execution control is not provided at operating system level, but at application-level through the use of continuations. This choice makes the framework flexible and easily portable across heterogeneous programming environments with support of continuations since there is no direct dependence with the operating system. The proposed framework represents a programming platform for developing and experimenting with service scheduling policies in different service-oriented applications.

References

1. Foster, I., Kesselman, C., Nick, J., Tuecke, S.: The physiology of the grid: An open grid service architecture for distributed system integration. Technical report Open Grid Service Infrastructure WG (2002)
2. Wooldridge, M.: Engineering the computational economy. In: Proceedings of the Information Society Technologies Conference (IST–2000), Nice, France (2000)
3. Buyya, R., Abramson, D., Giddy, J.: An economy driven resource management architecture for global computational power grids. In: Proceedings of The 2000 International Conference on Parallel and Distributed Processing Techniques and Applications (PDPTA 2000), Las Vegas, USA (2000)
4. Buyya, R., Giddy, J., Abramson, D.: An evaluation of economy–based resource trading and scheduling on computational power grids for parameter sweep applications. In: Proceedings of The Second Workshop on Active Middleware Services (AMS 2000). In conjuction with Ninth IEEE International Symposium on High Performance, Pittsburgh, USA (2000)
5. Foster, I., Roy, A., Sander, V.: A quality of service architecture that combines resource reservation and application adaptation. In: Proceedings of the 8th International Workshop on Quality of Service (IWQOS 2000), Pittsburgh, USA (2000) 181–188
6. Friedman, D.P., Haynes, C.T., Kohlbecker, E.E.: Programming with continuations. In: Program Transformation and Programming Environments, pp. 263–274. Springer, Heidelberg (1984)
7. Foster, I., Kishimoto, H., Savva, A., Berry, D., Djaoui, A., Grimshaw, A., Horn, B., Maciel, F., Siebenlist, F., Subramaniam, R., Treadwel, J., Reich, J.V.: The open grid services architecture, version 1.0. Technical report, Global Grid Forum Informational Document (2005)

8. Czajkowski, K., Foster, I.T., Kesselman, C., Sander, V., Tuecke, S.: Snap: A protocol for negotiating service level agreements and coordinating resource management in distributed systems. In: Feitelson, D.G., Rudolph, L., Schwiegelshohn, U. (eds.) JSSPP 2002. LNCS, vol. 2537, pp. 153–183. Springer, Heidelberg (2002)
9. Wäldrich, O., Wieder, P., Ziegler, W.: A meta-scheduling service for co-allocating arbitrary types of resources. In: Wyrzykowski, R., Dongarra, J.J., Meyer, N., Waśniewski, J. (eds.) PPAM 2005. LNCS, vol. 3911, pp. 782–791. Springer, Heidelberg (2006)
10. Vadhiyar, S., Dongarra, J.: A metascheduler for the grid. In: Proceedings of the 11th IEEE Symposium on High-Performance Distributed Computing, IEEE Computer Society, Los Alamitos (2002)
11. Di Napoli, C., Mango Furnari, M.: A continuation–based distributed lisp system. In: Proceedings of the First International Conference on Massively Parallel Computing Systems, pp. 523–527. IEEE Computer Society Press, Los Alamitos (1994)
12. Tismer, C.: Stackless python (2007), http://www.stackless.com
13. SourceForge.net: Python web services (2007),
    http://pywebsvcs.sourceforge.net
14. Newhouse, T., Pasquale, J.: A user-level framework for scheduling within service execution environments. In: Proceedings of the 2004 IEEE International Conference on Services Computing (SCC '04), pp. 311–318. IEEE Computer Society, Washington, DC (2004)
15. The Apache Software Foundation: Apache web services project - axis (2007),
    http://ws.apache.org/axis
16. IBM developerWorks: WebSphere (2007),
    http://www-128.ibm.com/developerworks/websphere
17. Johnson, R.E., Foote, B.: Designing reusable classes. Journal of Object-Oriented Programming 1(2), 22–35 (1988)
18. Fowler, M.: Inversion of control containers and the dependency injection pattern (2004), http://www.martinfowler.com/articles/injection.html
19. Booth, D., Liu, C.K.: Web services description language (wsdl) version 2.0 part 0 primer (2007),
    http://www.w3.org/TR/2007/PR-wsdl20-primer-20070523
20. Adams, H.: Asynchronous operations and web services, part 2 (2002),
    http://www-128.ibm.com/developerworks/library/ws-asynch2/index.html
21. Swenson, K., Ricker, J.: Asynchronous web service protocol (2002),
    http://xml.coverpages.org/AWSP-Draft20020405.pdf
22. Sun Developer Network: Developing asynchronous web services with java message service in sun java studio enterprise 7 (2005), http://developers.sun.com/prodtech/javatools/jsenterprise/reference/te chart/jse7/asynch.html
23. Platform: Open source metascheduler for virtual organizations with the community scheduler framework (csf). Technical report (2007), http://www.cs.virginia.edu/~grimshaw/CS851-2004/Platform/CSF_architecture.pdf
24. Open portable batch system (2007), http://www.openpbs.org
25. Load sharing facility (2007), http://www.platform.com

On Business Grid Demands and Approaches

Carsten Franke, Adolf Hohl, Philip Robinson, and Bernd Scheuermann

SAP Research CEC Belfast, Karlsruhe
{carsten.franke, adolf.hohl, philip.robinson, bernd.scheuermann}@sap.com

Abstract. This paper addresses necessary modification and extensions to existing Grid Computing approaches in order to meet modern business demand. Grid Computing has been traditionally used to solve large scientific problems, focussing more on accumulative use of computing power and processing large input and output files, typical for many scientific problems. Nowadays businesses have increasing computational demands, such that Grid technologies are of interest. However, the existing business requirements introduce new constraints on the design, configuration and operation of the underlying systems, including availability of resources, performance, monitoring aspects, security and isolation issues. This paper addresses the existing Grid Computing capabilities, discussing the additional demands in detail. This results in a suggestion of problem areas that must be investigated and corresponding technologies that should be used within future Business Grid systems.

Keywords: Grid Size, Software Landscapes, Application Data, Execution Characteristics, Autonomy, Service Level Agreements.

1 Introduction

The paradigm of Metacomputing [6] and later Grid Computing originated in the early 1990s and refers to the coupling of geographically dispersed computers, storage systems, scientific instruments etc. [13,10,12]. This enabled the execution of a wide range of scientific applications like e.g. large scale simulations, collaborative engineering or computer-aided instrumentation [6,7].

In the majority of businesses, the ability for their organization and technical infrastructure to adapt to dynamically changing business environments has become a key component in their success. Many businesses are searching for proven, technological solutions that enable them to execute with these new levels of adaptablity. In the world of large-scale scientific problem solving, similar adaptablity challenges have been faced and already addressed by allocating computational jobs to aggregations of distributed nodes. Although this seems a logical avenue to pursue for solving the business agility problem, such solutions are still not fully compliant with the operational constraints imposed by a business environment. The term *Business Grids* is hence introduced to distinguish the effort towards achieving this compliance in comparison to what currently supports scientific problem-solving.

D.J. Veit and J. Altmann (Eds.): GECON 2007, LNCS 4685, pp. 124–135, 2007.

We consider the current approaches to supporting Scientific Grids as having the ultimate goal of aggregating as much computational power as possible. Examples of toolkits and technologies that enable this goal are Globus [11], GRIA [24], and gLite [8]. These approaches can be used as foundations for building Business Grids but, due to the business-related constraints that we explain, there are some hindrances to direct usage.

This paper introduces some of the major requirements on Business Grids and discusses several technologies and concepts that might be used to overcome the current limitations of Scientific Grids. Note that the paper is not intended to describe particular business models, pricing model, market economies etc., as these are business-domain specific [9,25]. Moreover, such models can only be realized when the necessary technologies are in place. In our opinion, this is not the case, such that true evaluation of these models depends on the right technical frameworks and infrastructure being in place.

The remainder of this paper is structured as follows: In Section 2, we introduce and compare the characteristics of Scientific Grids and Business Grids. Afterwards, the extended, business-driven requirements for using Grid Computing are discussed in Section 3. Section 4 reviews some related work and technologies, deriving the challenges that need to be solved to fulfill the business demands. In Section 5, we identify key technologies and steps that may pave the way towards Business Grids. Finally, we summarize and conclude the paper in Section 6.

2   Scientific Grids and Business Grids

Within this section we elaborate on the applications that are executed on Scientific Grids today and the applications intended for hosting and execution in a grid-like infrastructure we call Business Grids. Note that the term "Scientific Grid" is meant to refer to Grid computations for scientific projects like e.g. the Large Hadron Collider (LHC)[4] or SETI@home [1], in the context of this paper.

2.1   Scientific Grids

A Scientific Grid is considered as a computing model for parallel processing across heterogeneous resources belonging to multiple geographically dispersed administrative domains. The computing problems considered require large a-mounts of either computing time or data, such that their reduction into several small parallel processes, with only little inter-process communication and a limited execution time (as opposed e.g. to interactive applications), sees computational improvements through parallelization. The majority of Scientific Grid computations can be characterized as stateless batch jobs mainly performing file-based input and output operations. Such jobs can be deployed in a relatively short time as they are submitted in a self-contained manner along with all input data files and executables required. The jobs usually do not depend on locally available license files or user interactions and are therefore highly mobile.

Data management in Scientific Grids focuses on data modeling, data movement, and handling of distributed and replicated files. Typically, data is modeled in comparatively flat data structures, and data access is performed in a non-transactional fashion. Resource management puts the main emphasis on high throughput and high resource utilization whereas low response times are often considered as less important. Security issues are commonly of minor interest since confidentiality and integrity of data are not critical in real-time, and computation resources and networks are assigned exclusively to scientific project members.

2.2   Business Grids

For the economic success of companies, it becomes more and more essential to be able to dynamically adapt to new business demands, knowledge and constraints. Consider a simple scenario for business agility where customer or application transaction load needs to be shifted or redirected to different application service providers, given variations in costs, availabilities and reputations. As a step towards such levels of flexibility, several companies host computing infrastructures for dedicated business tasks, billing on a per-use basis, labeling them as *e-business on demand, business on demand, utility computing* and others. However, such solutions do not support the levels of management automation required in order to respond effectively to dynamically changing business needs. Nevertheless, direct deployment of business applications to an existing scientific Grid infrastructure is still insufficient and too risky. Firstly, Business Grids should be able to restrict the sites providing resources to build the execution environment, requiring more complex resource and provider selection logic. One of the main reasons is that the participating parties need to have standing, legal contracts with each other regarding resource usage. In addition, they need to beware of global compliance concerns that could have damaging effects if breached as a result of transferring data to different sites. Furthermore, in order to obtain maximum flexibility that reflects the way in which ambitious businesses operate, technical security mechanisms with no assumed pre-existing trust relationships or anchors have to be assumed. Determining and initializing these on-demand still adds to the overhead of agility and requires more attention.

In contrast to Scientific Grids, which mainly interchange and process large, flat data files, Business Grids data processing must assume the existence of large relational databases, given the legacy of business systems. In such cases it is not possible to hide the transport of data to the executing node by pre-fetching all data before the job execution starts. In Business Grids, small sets of data must be loaded and stored from a database with a random access pattern. This does not only effect the execution time of the application itself, but also the scheduling decisions as the node that executes a certain application cannot be allocated too far from the database. Depending on its size the database remains static at a certain location. Additionally, the main applications within Business Grids are interactive, where interactions take place within open and stateful sessions, small data packets are exchanged frequently with the database, and no delays in data transmissions can be tolerated. Application components can be migrated but the interactive session must remain open. Even so, security levels of business applications are high in both the online and offline modes, in comparison with applications in Scientific Grids. The execution of the application itself must be secured and also all the related data before, during, and after the processing. Such security constraints also limit the allocation possibilities and the mobility of data.

If these business-related requirements were to be fulfilled, Business Grids could be used in various scenarios. For example, data centers could make high use of Business Grid technologies in order to increase the in-house adaptability to changing customer demands. The Grid technologies would be used as mechanisms for dynamic in- and outsourcing of computational capacity, sizing and re-sizing the deployment landscape, ensuring that all given business constraints are fulfilled. Additionally, the resources from different providers could be logically bound and identified as a new, collective functional capability that did not previously exist. This is today known as a Virtual Organization (VO) [10]. However, there remains a gap between the technical resource-sharing notion of a VO and the corporate management-sharing notion of business, which needs to be reconciled before Business Grids come into full existence.

3   Business Grids Requirements

In this section, we elaborate on the increasing demand for Grid technologies in business scenarios followed by a discussion of the various differences that upcoming Grid technologies must address in order to be compliant with the business demand. A comparison of Business and Scientific Grids is provided using identified problem areas, including resource composition, software landscape, application data, execution characteristics, execution characteristics, and service level agreements.

3.1   Grid Size

A fundamental difference between Business and Scientific Grids considers the scope and distribution of the computational nodes that form the infrastructure. Business Grids have been motivated by needs to serve single or very few administrative domains. They therefore have very closed assumptions about where data and applications are physically located, executed and managed. Scientific Grids are nevertheless developed with open-world assumptions and seek to acquire more and more resources without the hard constraints on the administrative domains involved [3]. This adds an additional constraint on the sizing and distribution properties of Business Grids that is currently not a concern for Scientific Grids. Furthermore, the structure of an organization places constraints on the topology and pricing model for resource distribution and usage.

3.2   Software Landscape

Business applications exhibit characteristics that differ from scientific or high performance computing applications. Since business applications implement both, standardized and individual business processes, they vary from enterprise to enterprise. They have to react on unschedulable/spontaneous events, triggered by user input or connected infrastructure components (e.g. logistic supporting RFID sensor infrastructure) as well as scheduled workload (e.g. payroll accounting in components supporting human resource tasks). From a birds eye perspective business solutions are deployed in a three tier architecture: The business application itself providing the business logic and user interfaces. This business logic is executed on a middleware platform (application server), with the third component in this architecture being a database. Since the transactions of almost all business applications access a central database, the database can be considered as a major medium for information exchange. Deploying such an infrastructure requires an individual configuration of each system and differs in the complexity to a Scientific Grid application where one and the same software has to be deployed multiple times. The size of the software which has to be deployed varies from a few to several hundred GB.

3.3   Application Data

Due to business and legal reasons, data and information of enterprises need to be accessible and readable for several years. Thus a lot of data and aggregated information has to be stored in a structured way. The size of used databases rises up to several TB. However small sets of data in the database are accessed frequently in a non-predictive manner. The implementation of sensor-triggered solutions (e.g. in the field of logistics or mass production) increases the demands on data management due to frequent transactional data access.

The application data has to be consistent and reflect the modeled reality. To achieve this, locks to subsets of data are applied and synchronous logs are written to guarantee this property. In contrast to the high performance computing, a relaxed consistency model is sufficient and computations are mostly performed independently. Usually, the problem space can be structured in an easy way and is divided into several parts. This is then sent to computation nodes. After finishing the computation, the result is returned and composed with others.

3.4   Execution Characteristics

Another main difference between Scientific Grids and Business Grids concerns application execution [16]. As already described, business applications are often interactive and frequently access databases. Such applications often have a very long runtime (up to years) in comparison to single, batch, scientific jobs. Furthermore, the deployment of these applications is rather long compared with the majority of scientific applications as it normally includes the installation and configuration of an entire multi-tier system including application servers and databases with up to several Gigabytes of data.

The comparatively longer execution times lead to the demand for mechanisms to migrate running business applications dynamically, as it is very likely that the underlying infrastructure must be maintained during the application runtime. However, the mobility of business application is limited due to the fact that each application and the corresponding data often has very strong security requirements and therefore is critical for the business itself. In contrast, the mobility of data is often not restricted within Scientific Grids.

Additionally, the business applications are in most cases split into several components and layers with separate executables. This also leads to quite complex start procedures for business applications. Here, the various dependencies between the operating system, the middleware, the data, and the application must be incorporated. In contrast, the majority of scientific applications consist of a single executable that can be started independently.

The majority of business applications are stateful, especially when databases are available for transaction data and transaction persistence. Thus, if cases of failures a simple restart of the application is not possible. This motivates the usage of concepts to either duplicate the different application executions or to develop methods that allow a very precise checkpointing of the applications such that states are not lost.

3.5   Autonomy

As the business processes are changing quicker than ever before, the underlying information technology must be able to adapt and to follow these changes. In the past, the required system modifications were mainly performed by human beings. This is expensive and also often not quick enough. Thus, business solutions drive the demand to reduce the amount of work that has to be spent in the management of the Grid. Hence, self-provisioning of applications and self-management of the Grid infrastructure are two major longer term goals, see Franke et al. [14]. Furthermore, self-healing strategies need to be developed that enable stateful applications to recover after failures. Additionally, business usage scenarios require the ability of self-optimization on different levels, like the application level, the middleware level, and the hardware level. All the above-mentioned requirements are in contrast to the existing Scientific Grid approaches. In most cases, restarting an application after a failure is sufficient in theses cases.

3.6   Service Level Agreements

One of the main differences between the scientific and the business usage of Grid technologies is the purpose itself. In the scientific environment, the main goal can be described as getting as much computing power as possible with a reasonable amount of management tasks for the participating scientific institutes. In the business context, Grids are used to flexibly execute applications that are provided to customers paying for this kind of service. Thus, one major goal is to fulfill the business requirements with all the corresponding constraints by automating as much as possible.

All the constraints of the application execution are defined within a Service Level Agreement (SLA) which reflects a legal contract between the service provider and the customer. Usually, an SLA is a bidirectional contract that specifies in detail what service the provider must deliver and which rules the customer agrees to follow. SLAs also describe the agreed reliability of the service, the billing and other business-critical issues. Furthermore, the security constraints are detailed within SLAs by specifying the kind of isolation that the service provider must deliver (user isolation, application isolation, performance isolation). As the business is directly connected to such SLAs, a very sophisticated SLA management is required. Therefore, industry demands Business Grids to vertically integrate an automated SLA management across all layers of the execution stack.

4   Related Work

There are various technologies and concepts in existence and development that might contribute to the generation of Business Grids. A review of some of these possible concepts shows that there are important strategic and technical challenges that need to be solved before Business Grids can be fully realized.

Information Systems Outsourcing: Information Systems Outsourcing is the contracting of various systems to external information system providers for operational and/or maintenance purposes [18,15]. The simple argument for outsourcing is that the customer focuses on their business domain, while the provider relieves them of significant overhead and risks associated with maintaining large-scale, complex technology. From the customer's perspective they run their business applications without hiring additional, specialists staff or worrying about financing additional utility costs. From the provider's perspective they offer specialists services at a price that is less than the potential hiring, training and utilities costs faced by their customers, while having control over how they manage and reuse their resources. These sorts of arguments are not particularly of relevance for Business Grids for three reasons: (1) Business Grids seek to keep the customer in control of resource management, (2) customers and users of a Business Grid determine the levels of infrastructure transparency they require, and (3) Business Grids allow to enforce that providers fully respect the resource selection, deployment and scheduling constraints of customers.

Service-Oriented Computing: Service-oriented Computing provides software integration constructs and mechanisms that allow distributed processing and storage units to be flexibly linked together in a loosely-coupled manner [20]. That is, services maintain their autonomy and encapsulation and interact strictly using message passing, following the fundamental definition of a distributed system architecture. A service specification acts as descriptor and invocation interface for specific, well-defined and encapsulated functionality, which can be composed with others to form more complex systems and workflows. Depending on the protocols and accessibility of services, resultant systems can be distributed across multiple administrative domains, allowing dynamic discovery and replacement. Many of the technical capabilities and facilities for Service-Oriented Computing can be applied in the context of the Business Grid. However, the description concept of service oriented computing only allows for the use of static attributes (service names, functions and categories) during resource discovery, whereas these attributes need to be more dynamic and varied for the purposes of Business Grids.

Utility Computing: Utility computing [21,22] is rather a business concept than a technological advance, where users of a large-scale computing resource pay only for the computational power, storage and software that they use, similar to the way in which electricity, telephones and water are charged. Users therefore subscribe to computational utility providers and agree on their terms of usage, which may be measured based on volume, time or quality. While the Business Grid requires a similar model of subscription, pricing and resource reservation, the underlying computational infrastructure must be leveraged to flexibly adapt to dynamically changing business requirements.

Virtual Clusters: The term *Cluster Computing* is often misused as a synonym for Grid Computing. While clusters are typically bound over short-range, high-speed communication links and controlled by a centralized scheduler, Grid nodes are interconnected over possibly wide-area, variable-speed networks, with sophisticated scheduling methods for the actual resource allocation over time. The applications deployed on clusters therefore differ from those enacted on a Grid. Secondly, a Grid can be composed of several clusters, making it a larger-scale and potentially more complex computational infrastructure for supporting distributed applications. Therefore, the hosting of business applications on a cluster does not constitute the concept of Business Grids that we propose. Virtual Clusters provide an abstraction for regulating computational resource usage for different groups of users. Selected nodes of a cluster are reserved for different users or groups by the specification of resource sharing rules, such that the groups are not aware of the full power available. From our perspective, Virtual Clusters (VCs) can be seen as complimentary to the Virtual Organization (VO) concept, where VCs technically refer to how resources are bound across domains, while VOs provide a management model. Business Grids also require mechanisms derived from Virtual Clusters for resource regulation, yet the means by which they are enforced need to be more sophisticated than message interception using reference monitors.

5   Steps on the Way to Business Grids

This section describes different approaches and technologies that can be used in order to make a Grid ready for the execution of business applications.

First the technologies which we think are of major interest are briefly introduced. Since there is no one-to-one relationship of the discovered requirements and the proposed technologie their concrete exploitation is mentioned in the following.

– Virtual Machines: Virtualization of a machine such as [2,17] multiplexes multiple virtual instances to a physical machine. It can improve utilization but more importantly provide flexibility in the management of systems. From a birds eye perspective most technologies provide the same primitives independent of their technological implementation. Process checkpointing technologies can provide similar primitives too and are also of interest.
– Distributed Filesystem: Distributed filesystems can from a birds eye perspective seen as a multiple instantiation of network filesystems such as NFS. Information is accessed through multiple pathes which promise a higher throughput when multiple clients accessing data. Usually data and its metadata is stored in separately, which can improove metadata access. A replication policy controls which and where data is replicated to. Representatives are CEPH, PVFS [19,5].
– Deployment Infrastructure: Deployment infrastructure focuses on the rollout and management of software and its configuration. Provided a description of the interdependencies of the software, it can be deployed and started in the right order. One representative implementation is SmartFrog [23].

A first solution proposal to the discovered deficiencies is described next. Table 1 shows the technical mechanisms which are considered and how they are exploited with respect to the various *problem areas* introduced in Section 3.

Table 1. Assessment of the various technical mechanisms with respect to Grids aspects

Problem Areas	Virtual Machines	Distributed File Systems	Deployment to Infrastructure
Grid Size	-	-	Affects
Software Landscape	Affects	Affects	Affects
Application Data	Affects	No	Affects
Application Execution	Affects	No	Affects
Autonomy	No	No	No
SLA	Affects	No	Affects

Software Landscape: The deployment routines for deploying a business application have to be aware of system dependencies. Frameworks such as SmartFrog are promising to provide a solution to this. Nevertheless, the installation procedure can be very time consuming. Instead of using traditional software installers the use of virtualization technologies can accelerate this task by deploying ready to use disk and virtual machine images.

Application Data: The ability to deploy the database at a location different from the other parts of the business application can be supported by the use of advanced caching mechanisms to compensate network latency above local networks. Access to often used pages to the database file are cached to local disks in the vicinity of the application and thus replicated incrementally. A distributed file system which replicates to the most used deployment sites may also provide a solution to the migration issue of huge data files. Although there are a lot of distributed file systems available their suitablity for database typical workload has to be investigated.

Application Execution: The constraints for the application execution can be fulfilled by the combined usage of several new technologies and the slight adaptation of existing tools. In general, virtualization techniques can be used to enable checkpointing of existing legacy software with only little need to modify these software solutions. Furthermore, the virtualization techniques would enable the migration of applications during the runtime by migrating the virtual execution container. This would solve the problem of infrastructure maintainance for applications with long runtimes. The problem of database connectivity to the business application can only be solved by an adapted deployment. This deployment must take care that the application and the database are located close enough such that the database access from the application has a reasonable response time. This response time should be specified in the corresponding SLA. Additionally, the deployment of business applications must incorporate the various application layer and all corresponding constraints. Note that checkpointing or migration can also be performed on the application directly (see e.g. BLCR, CHPOX, CRAK, HPC4U, UCLiK). However, the problem with such approaches is that entire goups of processes belonging to the application along with all open communications and files have to be captured and tranferred to a compatible hardware architecture.

Autonomy: This issue is not addressed in existing Grid approaches. Thus, the Grid management itself must adapt functionalities that enable a kind of self-management and self-adaptability. All the related issues are at the moment open research topics.

Service Level Agreements: This topic is addressed in several Grid approaches. However, the current state is primarily dominated by scientific requirements and mainly focuses only on the physical infrastructure. Thus, current approaches mainly specify hardware characteristics. However, the requirements of businesses are different in the sense that the higher level SLAs between the service providers and the customers must be broken down onto the infrastructure. To this end, the complex software landscape and virtualization technologies must be incorporated. Furthermore, the SLAs must consist of functional and non-functional service specification. Theses issues are also open research topics that must be addressed soon in order to make Grid technologies applicable for businesses.

6 Conclusion

Business applications were developed and further improved for decades on mainframes and client/server infrastructures. As the businesses need to react more dynamically than ever before, the underlying information technology also needs to adapt dynamically to changing environments. Thus, the application of Grid technologies for business solutions seems reasonable.

However, making all these applications fully aware of a Grid Computing infrastructure is not feasible. Therefore, this paper starts to find a greatest common divisor between what is provided by current Grid technologies and related approaches and the business application demands.

In this paper, we pointed out differences between the Grid and the business application domains as well as technical deficiencies of current Grid approaches from the business perspective. Furthermore, technical mechanisms are proposed to address most of these issues. One of the major interests is the investigation of technologies which alleviate the binding of applications to dedicated physical resources. In general, business applications could benefit most from Grid infrastructures if these applications could dynamically be deployed in a distributed manner.

Of course, the business context restricts the flexibility during the distribution which must be incorporated at the deployment time. Ongoing efforts to build business application on a service oriented architecture provide the technical precondition to further increase the potential benefit of using Grids in business environments. Services allow for a finer-grained deployment on Grid nodes than traditional applications thereby increasing the flexibility of business solutions and leading to a higher adaptability of the Grid.

Acknowledgments. We would like to thank the support from the European Commission under IST program #FP6-033576.

References

1. Anderson, D.P., Cobb, J., Korpela, E., Lebofsky, M., Werthimer, D.: Seti@home: an experiment in public-resource computing. Communications of the ACM 45(11), 56–61 (2002)
2. Barham, P., Dragovic, B., Fraser, K., Hand, S., Harris, T., Ho, A., Neugebauer, R., Pratt, I., Warfield, A.: Xen and the art of virtualization. In: SOSP '03: Proceedings of the nineteenth ACM symposium on Operating systems principles, pp. 164–177. ACM Press, New York (2003)
3. Brune, M., Gehring, J., Keller, A., Reinefeld, A.: Managing clusters of geographically distributed high-performance computers. Concurrency - Practice and Experience 11(15), 887–911 (1999)
4. Bunn, J.J., Newman, H., McKee, S., Foster, D.G., Cavanaugh, R., Hughes-Jones, R.: Bandwidth challenge - high speed data gathering, distribution and analysis for physics discoveries at the large hadron collider. In: Löwe, W., Südholt, M. (eds.) SC 2006. LNCS, vol. 4089, p. 241. Springer, Heidelberg (2006)

5. Carns, P., Walter, H., Ross, R., Thakur, R.: Pvfs: A parallel file system for linux clusters. In: Proceedings of the 4th Annual Linux Showcase and Conference, Atlanta, GA, USENIX Association, pp. 317–327 (2000)
6. Catlett, C., Smarr, L.: Metacomputing. Communications of the ACM 35(6), 44–52 (1992)
7. DeFanti, T., Foster, I., Papka, M., Stevens, R., Kuhfuss, T.: Overview of the i-way: Wide area visual supercomputing. International Journal of Supercomputer Applications 10(2), 123–130 (1996)
8. Dvořák, F., Kouril, D., Krenek, A., Matyska, L., Mulac, M., Pospíšil, J., Ruda, M., Salvet, Z., Sitera, J., Vocu, M.: glite job provenance. In: Moreau, L., Foster, I. (eds.) IPAW 2006. LNCS, vol. 4145, pp. 246–253. Springer, Heidelberg (2006)
9. Ernemann, C., Yahyapour, R.: Grid Resource Management - State of the Art and Future Trends. In: chapter Applying Economic Scheduling Methods to Grid Environments, pp. 491–506. Kluwer Academic Publishers, Dordrecht (2003)
10. Foster, I.: What is the grid? a three point checklist. GRIDtoday, 1(6) (2002)
11. Foster, I., Kesselman, C.: The Globus project: a status report. Future Generation Computer Systems 15(5–6), 607–621 (1999)
12. Foster, I., Kesselman, C.: The Grid: Blueprint for a New Computing Infrastructure. Morgan Kaufmann Publishers, San Francisco (1999)
13. Foster, I., Kesselmann, C.: The Grid: Blueprint for a New Computing Infrastructure. Morgan Kaufmann Publishers, San Francisco (1998)
14. Franke, C., Theilmann, W., Zhang, Y., Sterritt, R.: Towards the autonomic business grid. In: Fourth IEEE International Workshop on Engineering of Autonomic and Autonomous Systems (EASe'07), pp. 107–112 (2007)
15. Goles, T.: Vendor capabilities and outsourcing success: A resource-based view. Wirtschaftsinformatik 45(2), 199–206 (2003)
16. Kenyon, C., Cheliotis, G.: Grid Resource Management - State of the Art and Future Trends. In: chapter Grid Resource Commercialization - Economic Engineering and Delivery Scenarios, pp. 465–478. Kluwer Academic Publishers, Dordrecht (2003)
17. Kolyshkin, K.: Linux virtualization (2005)
18. Nam, K., Rajagopalan, S., Raghav Rao, H., Chaudhury, A.: A two-level investigation of information systems outsourcing. Commun. ACM 39(7), 36–44 (1996)
19. Olson, C., Miller, E.L.: Secure capabilities for a petabyte-scale object-based distributed file system. In: StorageSS '05: Proceedings of the 2005 ACM workshop on Storage security and survivability, pp. 64–73. ACM Press, New York (2005)
20. Papazoglou, M.P., Georgakopoulos, D.: Introduction. Commun. ACM 46(10), 24–28 (2006)
21. Rappa, M.A.: The utility business model and the future of computing services. IBM Syst. J. 43(1), 32–42 (2004)
22. Ross, J.W., Westerman, G.: Preparing for utility computing: The role of it architecture and relationship management. IBM Systems Journal 43(1), 5–19 (2004)
23. Sabharwal, R.: Grid infrastructure deployment using smartfrog technology. In: ICNS, page 73 (2006)
24. Surridge, M., Taylor, S., De Roure, D., Zaluska, E.: Experiences with gria industrial applications on a web services grid. In: E-SCIENCE '05: Proceedings of the First International Conference on e-Science and Grid Computing, pp. 98–105. IEEE Computer Society, Washington, DC (2005)
25. Yeo, C.S., Buyya, R.: A taxonomy of market-based resource management systems for utility-driven cluster computing. Software: Practice and Experience 36(13), 1381–1419 (2006)

Enabling the Simulation of Service-Oriented Computing and Provisioning Policies for Autonomic Utility Grids

Marcos Dias de Assunção[1,2], Werner Streitberger[3],
Torsten Eymann[3], and Rajkumar Buyya[1]

[1] Grid Computing and Distributed Systems (GRIDS) Laboratory
[2] NICTA Victoria Research Laboratory
The University of Melbourne, Australia
{marcosd, raj}@csse.unimelb.edu.au
[3] Chair for Information Systems Management
University of Bayreuth, Germany
{streitberger, eymann}@uni-bayreuth.de

Abstract. There are key challenges in utility computing environments such as the provisioning, orchestration and allocation of resources to services. In these environments, providers need to decide how resources are allocated to service applications according to their workloads in order to guarantee the Quality of Service (QoS) required by customers. Autonomic computing inspired mechanisms are appealing to enable self-organising resource allocation and provisioning. However, these mechanisms are difficult to evaluate in practice either because of the lack of a real test bed or the difficulty in replicating experimental results. This work thus describes a service framework for a Grid simulator. This framework allows the modelling and evaluation of the provisioning and negotiation of services and resources. We also discuss experimental results that demonstrate the usefulness of this framework for the simulation of a decentralised and self-organising economic model for service and resource negotiation termed *Catallaxy*.

Keywords: Resource provisioning, Grid computing, utility computing, simulation framework.

1 Introduction

Service-Oriented Architectures (SOAs) underlie several Grid initiatives and reflect the current Grid computing infrastructure, where participants offer and request application services. A SOA defines standard interfaces and protocols that enable the encapsulation of resources of different complexity and value as services that clients access without having knowledge of their internal workings [1].

In current utility computing environments, resource providers host services and provide the tools needed by scientists and companies to expose the core functionalities of their research or business as services that are subsequently used by clients or collaborators; providers offer their resources generally in a

D.J. Veit and J. Altmann (Eds.): GECON 2007, LNCS 4685, pp. 136–149, 2007.

pay-as-you-go manner. Virtualisation technology offers powerful resource management mechanisms for these environments by enabling performance isolation,
migration, suspension and resumption of Virtual Machines (VMs). One key issue, however, is the provisioning, orchestration and allocation of resources to
services. Providers need to decide how resources are allocated to service applications according to their workloads in order to guarantee the QoS expected by
their customers. Autonomic computing [2] inspired mechanisms and policies are
appealing to enable self-organising allocation of resources to services, as well as
for service provisioning and negotiation [3, 4].

However, it is challenging to design and evaluate practical allocation policies
that permit utility computing environments to self-manage and adjust resource
allocations according to the provisioning decisions of the offered services. Moreover, it is a challenge to evaluate these policies and negotiation strategies either
due to the difficulty of replicating experiments or a lack of a real testbed.

The modelling and evaluation of these mechanisms and related policies can
be augmented by the use of simulators. However, current simulation tools focus
on issues related to resource modelling and allocation assuming in general a job
abstraction. The existing Grid simulation toolkits do not provide the features
needed to model and simulate services, their placement on resources, their workloads and provisioning policies let aside the abstraction of containers or VMs.
In this work, we present a framework that allows the modelling, simulation
and evaluation of mechanisms and policies for service provisioning, negotiation
and resource management. This framework supports the simulation of service-
oriented applications, and considers service dependencies, for different domains
including high-performance, on-demand and utility computing. We demonstrate
the usefulness of our framework by modelling and simulating an Application
Layer Network (ALN) and an economic model termed *Catallaxy* for service and
resource negotiation.

The rest of this paper is organised as follows. Section 2 presents background
and related work. Section 3 describes the service framework. In Section 4, we
present the design of a decentralised economic bargaining model for ALNs (i.e.
the Catallaxy). Section 5 presents the performance evaluation results and finally,
Section 6 concludes the paper.

2   Background and Related Work

In order to demonstrate the mechanisms and policies that we want to model
and simulate, we consider a utility data centre that hosts service applications,
and provides resources on demand to its customers' business applications (see
Fig. 1) [5]. The centre is composed of a pool of physical resources that are managed by server virtualisation technology [6]. The services offered to customers
run on Application Environments (AEs) within the resource pool, which are
isolated from one another. An AE is a set of virtual resources (i.e. containers
or VMs). The resource arbitrator allocates resources to each AE according to
the resource allocation policies in order to meet the required performance and

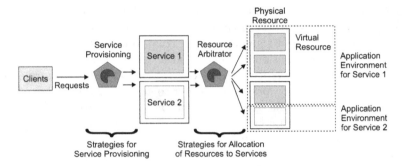

Fig. 1. Abstract view of a utility data centre

QoS. Customers can utilise services without the knowledge of the internal infrastructure of the resource layer and the resource allocation policies. However, customers and providers negotiate the Quality of Service (QoS) required, and customers want to have guarantees about the service delivery. These guarantees are stated in Service Level Agreements (SLAs). Service provisioning policies define how the service is provided in order to achieve the service levels stated in the SLA. In this case, the provider has to decide on how the service is provisioned.

The services have a workload that can vary. The number of requests to the hosted services and the expected QoS will guide the arbitrator on the resource allocation decisions. The arbitrator decides on the resources required by each service and if new resources have to be allocated to meet peak demands. A decoupling of service and resource layers allows one to model strategies for the placement of services on resources and resource orchestration. One can also evaluate distinct markets or mechanisms for service negotiation and resource allocation. Therefore, a provider has two policies: one that defines how a service is provisioned and another that specifies how resources are allocated.

The scenario above is an example; however, a simulation framework should be flexible enough to enable the modelling and simulation of varying scenarios. For instance, the ALN presented in this work follows a two-layer market model. In one layer, resource providers provide processing and storage resources. Service providers negotiate with resource providers to acquire capacity to host services. The second layer corresponds to the negotiation between service providers for the delivery of composite services. For example, a service provider can negotiate the access to several atomic services in the service market to deliver it as a bundle, or composite service, to its customers. Similar scenarios are considered in other utility computing strategies [7].

2.1 Related Work

Several Grid simulators allow the modelling and simulation of Grid resources and allocation policies; examples include OptorSim [8], SimGrid [9] and MicroGrid [10]. OptorSim is a discrete event simulator that follows the abstraction of data resources. It has been designed to model and evaluate the data transfer strategies for data Grids, and does not provide a service-oriented application model.

MicroGrid enables the emulation of Grid environments. A user can run his Grid application on an emulated environment, while the simulator intercepts the exchanged messages. Although it is possible to simulate service-oriented applications, MicroGrid does not provide a decoupling of the service and resource layers, which would enable the design and evaluation of different mechanisms for each layer.

SimGrid is a trace-based event simulator that provides a set of abstractions and functionalities to build simulators for several application domains. The core features can be used to model and evaluate parallel application scheduling on distributed computing platforms. SimGrid also provides emulation facilities for running distributed and parallel applications in an emulated Grid environment. SimGrid like the the other simulators, uses the abstraction of 'resources'.

GridSim [11] is a Grid simulation toolkit that enables the modelling of application composition, information services, and heterogeneous computational resources of variable performance. GridSim also provides an auction framework for the design and evaluation of auction protocols for Grid systems. With these features, it is possible to model and evaluate the scheduling of jobs on Grid resources and the impact of varying allocation policies. Similar to other simulators, GridSim enables the design and modelling of the resource layer.

In this work, we leverage the existing features of GridSim and provide a service framework that enables the modelling and evaluation of service provisioning policies, resource allocation policies and multiple economic mechanisms for service negotiation and resource management. GridSim, along with the extensions described here, provides means for evaluating autonomic computing systems, utility computing environments and utility Grids.

3   A Service Framework for GridSim

GridSim [11] is a discrete event simulator built on top of SimJava2 simulation package. A simulation in GridSim comprises of GridSim entities that communicate with one another by scheduling simulation events. Applications are modelled as jobs that are executed on Grid resources. A *Gridlet* represents a job and has parameters like job length expressed in Millions of Instructions (MIs), amount of CPUs required, among others. It is possible to model Grid resources of varying configurations, where the processing capability of the resource's CPUs is expressed in Millions of Instructions Per Second (MIPS). GridSim provides default resource allocation policies (e.g. space-shared, time-shared and space-shared supporting advance reservations), but the user can develop his own.

GridSim provides a hierarchical Grid Information Service (GIS) that can comprise of multiple regional GISs. At the start of the simulation, a Grid resource registers itself with a regional GIS. By default the Grid resource registers only its resource ID and indicates whether it supports advance reservation; however, the user can specify additional information to be provided to the GIS.

Based on the utility computing scenario described in Section 2, we design the framework considering two distinct stages: (i) the negotiation for and allocation of the resources to host services, and the negotiation for services and the required QoS; and (ii) the actual utilisation of the services and resources. The framework provides means for modelling service registries and discovery, service and resource negotiation as well as means for measuring the resource utilisation imposed by the services' workloads. A provider in this scenario has two policies: one that defines how a service is provisioned and one that defines how resources are allocated to a service. The allocations may change according to the service workloads.

Fig. 2 demonstrates the relationship between the main classes of the framework. The class *Provider* is a GridSim entity that implements the basic behaviour of a provider. A provider has characteristics represented by *ProviderCharacteristics*. The class *ProviderCharacteristics* contains a list of *Services* offered by the provider and other attributes like time zone, and the provisioning and acquisition policies utilised. *Service* corresponds to a service offered by the provider and has *ServiceAttributes* and a *ServiceRequirementList*. At the start of the simulation, the provider registers itself and the attributes of her services with a regional Grid Service Registry (GSR). *ServiceAttributes* include information like service cost, name and type. We implement service attributes as a distinct class for the sake of performance and minimisation of simulation events. *ServiceRequirements* correspond to atomic services or specific resources required to deliver the service to clients. For example, a provider may offer a service, but does not allocate resources to it until the service is required.

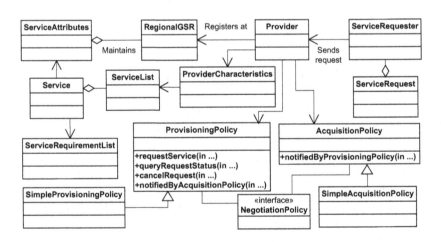

Fig. 2. Main classes of the framework

The Provider can engage in a market with clients for negotiating its resources. It can also participate in different markets with different mechanisms for negotiating and providing the resources necessary to host the services and satisfy the requests for a service. Both *ProvisioningPolicy* and *AcquisitionPolicy* implement the *NegotiationPolicy* interface. *NegotiationPolicy* defines the methods necessary to handle negotiations for service provisioning or resource allocation based on WS-Agreement. *ProvisioningPolicy* defines how a service is provisioned while the *AcquisitionPolicy* specifies the resource allocation. In other words, the *ProvisioningPolicy* defines how the provider manages the negotiation with clients for service provisioning and how it handles the resource requests. *Acquisition-Policy* specifies the provider's behaviour in negotiating with other providers for accessing the required services or resources. These services may be needed for composite services and the resources are required to host service applications. In many instances, provisioning and acquisition policies have to be synchronised or informed about one another decisions, as demonstrated by Grit *et al.* [12]. We provide methods that allow the policies to be synchronised.

Two examples of provisioning and acquisition policies are provided. In *SimpleProvisioningPolicy*, the provider accepts requests while the maximum number of instances for the service is not achieved. *SimpleAcquisitionPolicy* selects the first resource from the provider's resource pool to deal with the workload generated by the service requests. Although the Provider class can be extended, it is not necessary since it is possible to define different behaviours for a provider by extending the *ProvisioningPolicy* and *AcquisitionPolicy* classes to provide the strategies required.

The *ServiceRequester* class is a GridSim entity that queries services at a GSR and makes requests to providers. These queries can be performed by passing a filter to the GSR, which corresponds to specifying the parameters for a query. For example, the service requester can pass an object whose class extends *ServiceFilter* to select all the *ServiceAttributes* with a given service type and name. The GSR uses the filter to select and return a list of *ServiceAttributes* that match the given criteria.

A request for a service accepted by a provider generates a workload. The workload is composed of items that can be either requests for atomic services or *ServiceGridlets* that are sent to the resources allocated to the service. The *ServiceGridlet* class extends *Gridlet* by specifying additional parameters such as memory and storage required to fulfil the request. The values of these parameters for a service request can be estimated through profiling techniques, such as those described by Urgaonkar *et al.* [13], where a service application is examined in isolation and its workload is obtained by analysing the use of resources such as memory, CPU and disk. By following this model, it is possible to analyse the impact of different provisioning and acquisition decisions on resource utilisation.

4   Modelling the Catallaxy Scenario with GridSim

The CATNETS project investigates the use of an economic model, termed Catallaxy, for service negotiation and resource allocation in ALNs, such as Grids and P2P networks. Catallaxy is a decentralised self-organising economic model derived from Hayek's concept of spontaneous order [14]. The Catallaxy is based on the self-interested actions of participants, who try to maximise their own utility under incomplete information and bounded rationality. The goal of Catallaxy is to achieve a state of coordinated actions, through the bartering and communication of participants, to achieve a common goal that no single user has planned. Hayek's Catallaxy is the result of descriptive and qualitative research about economic decision-making of human participants. Its results are taken to construct ALN markets with software participants who reason about economic decisions using artificial intelligence.

The interdependencies between services and resources existing in ALNs are separated by creating two interrelated markets: a resource market for trading of computational and data resources; and a service market for trading services. This separation allows instances of a service to be hosted on different resources [15]. Fig. 3 shows the abstract model adopted by CATNETS. A Complex Service (CS) is a composite service, like a workflow, that requires the execution of other interdependent services, termed Basic Services (BSs). A CS is the entry point for the application layer network. The traded products on the service market, the BSs, are completely standardised and have a single attribute name. The name is a unique identifier whose intended semantics is shared among all complex service providers. Multiple instances of the same BS can co-exist in the ALN. For example, two or more basic service providers are allowed to provide a specific BS. The service market is used by Complex Service Providers (CSPs) to allocate BSs from Basic Service Providers (BSPs). BSPs are registered in a GSR. A CSP queries a GSR to receive a list of required trading partners (BSPs) able to provide the BS required. This list is ranked according to the BS offered price. The best BS offer is selected for the succeeding bargaining process. This discovery process is modelled using GSRs and discovery process offered by the simulation framework.

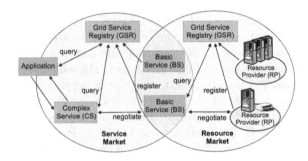

Fig. 3. The Catallaxy market model

After a successful negotiation in the service market, BSPs negotiate with Resource Providers (RPs) for the resources necessary to host services and serve the service requests. RPs utilise the existing resource management systems to allocate the necessary resources. RP offer resources in Resource Bundles (RBs). A resource bundle is described by a set of pairs of resource type and quantity. Every BS has an associated resource bundle. The bundle defines the type and quantity of resources needed for provisioning that service. In the CATNETS scenario, the resource bundle required for a BS is predefined for the sake of simplicity. In general, the model allows the use of any BS to resource bundle mapping function. In the resource market, the allocation process follows the service market. First, a BSP queries for RPs which are able to provide the specified resource bundle and ranks the received list of RPs according to the offered price. Second, the bargaining for the resource bundle is carried out. If the resource negotiation ends successfully, the BS is executed on the contracted resources from a RP.

To realise these two markets in GridSim, we implement provisioning and acquisition policies for the three kinds of providers (i.e. CSPs, BSPs and RPs). The providers differ in terms of the policies used for service and resource provisioning and acquisition. The execution of a market participant's policy for acquiring services or resources (i.e. *AcquisitionPolicy*) is shown in Algorithm 1 and that of a market participant's policy for service provisioning (i.e. *ProvisioningPolicy*) is depicted in Algorithm 2.

The most important part of the implemented policies is the utilised bidding strategy. This includes what a provider bids. The bid denotes the provider's valuation and reservation prices, i.e. the maximum price which an agent is willing to pay for the service and the minimum price an agent has for selling a BS or a RB respectively. The generation of the valuation is influenced by external factors

```
loop
   event ← wait for an event
   if event = message from provisioning policy then
     proposals ← ∅
     request_accepted ← the request ∈ event
     CFP ← create Call For Proposal (CFP) for request_accepted
     CFP
     proposals ← collect the proposals
     best ← select best proposal ∈ proposals
     start bargaining process
     outcome ← result of the bargaining process
     if outcome = success then
        inform other participants about the success
     end if
     apply learning algorithm
     notify provisioning policy about outcome
   end if
   if event = learning message then
     treat message received
     apply learning algorithm
   end if
end loop
```

Algorithm 1. Pseudo-code of the execution of an acquisition policy

```
loop
   event ← wait for an event
   if event = Call For Proposal then
      CFP ← get the Call For Proposal (CFP) ∈ event
      proposal ← formulate proposal for CFP
      reserve the resources
      send proposal
   end if
   if event = bargaining then
      start bargaining process
      outcome ← result of bargaining process
      if outcome = success then
         notify acquisition policy
         inform other participants about the success
      else
         release resources
      end if
      apply learning algorithm
   end if
   if event = reject proposal then
      release the resources
   end if
   if event = learning message then
      treat message received
      apply learning algorithm
   end if
end loop
```

Algorithm 2. Pseudo-code of a provisioning policy

such as the market price and the learning algorithm. For the formal model of the implemented strategy we refer to the work by Reinicke *et al.* [16].

The proposed realisation for the CATNETS markets is the usage of a bilateral negotiation protocol for exchanging bids in a point-to-point communication. Initially, both trading partners define a reservation price that reflects their estimation of the value of the good. For a buyer, this is the maximum price; for a seller it is a minimum price. The start price represents the negotiation starting point. By subsequent concessions, the opponents move closer to a compromise and a possible contract. Each opponent tries to maximise its own utility, which is the difference between the price of purchase and the reservation price. Thus, the buyer's and seller's policies converge to a trade-off point in an iterative way using the exchange of offers and counter-offers and successive concessions.

In the implementation in GridSim, CSs and BSs are modelled as *Services*. The service requirements of a CS define the BSs that are needed to deliver the CS. The service requirements of a BS define a minimum RB required to host the BS. The requirements of a BS j are represented by $BSR_j = (u_j, p_j, y_j, m_j, s_j)$, where u_j is the number of resources required; p_j represents the number of CPUs in each resource; y_j is the speed of the processors in MIPS; m_j is the amount of memory per resource; and s_j represents the storage capacity required.

A RP has a resource pool within which it creates Application Environments (AEs) with the resource configuration required by a BS. A RB corresponds to the resources offered by the RP. A RB i is represented by $RB_i = (u_i, p_i, y_i, m_i, s_i)$, where u_i is the number of resources in the bundle; p_i represents the number of CPUs in each resource; y_i is the speed of the processors in MIPS; m_i is the amount of memory per resource; and s_i represents the storage capacity per resource. A RP registers the RB with the GSR, which is viewed as a service by the BSP. That is, the RP provides a service that enables the BSP to acquire resources.

The negotiation for the resources needed by the BS starts after the negotiation for a BS is complete. The BSP searches for RPs that can provide a RB with the minimum amount of resources required. The BSP then starts the negotiation by sending a Call For Proposal (CFP) to the selected RPs. The BSP bargains with the RP that offers the best proposal. When the bargaining process ends, the RP allocates its resources to host the BS. Although a RP can divide its resource pool in various ways and change the allocations of AEs over time, in the CATNETS implementation, we consider that they are predetermined and do not change. The strategy used by a RP when it receives a CFP from a BSP during the negotiation of resources for a BS is summarised in Algorithm 3.

```
BSRⱼ ← get required resource bundle from the CFPⱼ
RBᵢ ← the resource bundle advertised
selected_resources ← ∅
booking_id ← 0
for all resource Rᵢ ∈ RBᵢ do
  if Rᵢ is not allocated then
    if pⱼ ≤ pᵢ and yⱼ ≤ yᵢ and mⱼ ≤ mᵢ and sⱼ ≤ sᵢ then
      selected_resources ← selected_resources ∪ Rᵢ
    end if
  end if
  if selected_resources = uⱼ then
    booking_id ← book(selected_resources)
    break for
  end if
end for
if booking_id = 0 then
  proposal ← create_proposal(selected_resources)
  send proposal
else
  reject CFPⱼ
end if
```

Algorithm 3. RP's strategy upon the arrival of a Call For Proposal (CFP) j

5   Performance Evaluation

We present experimental results that demonstrate how GridSim with the extensions discussed in this work can be used to model and evaluate service provisioning and resource allocation policies for service-oriented Grids and autonomic utility computing environments. The experiments particularly measure how the Catallaxy model, built on top of the discussed framework, coordinates the use of services and resources. We evaluate the allocation rate by identifying the number of service requests that are satisfied and the overhead imposed by the service and resource negotiations.

5.1   Experimental Scenario

We consider an environment in which RPs provide resource bundles and BSs require a particular resource bundle for a given time slot to host the service and execute the service workload. The experiments have been carried out considering a CS termed Workflow Service (WFS) that requires two BSs, namely Processing Service (PS) and Storage Service (SS). These two BSs, in turn, require a Processing Bundle (PB) and a Storage Bundle (SB) respectively. PB has the following configuration: $(p = 2, y = 1500MIPS, m = 1GB$ and $s = 2GB)$, while SB is given by: $(p = 1, y = 1500MIPS, m = 2GB$ and $s = 4GB)$.

We perform our experiments with varying numbers of RPs, BSPs and CSPs. The parameters used in the experiments are shown in Table 1. The values for PS Request Length (PSRL) and SS Request Length (SSRL) are given by $WSRL/2$ because we consider that WFS first requires processing and further stores the results of the processing activity. For simulating the workload of PS and SS and obtaining the final time of the service utilisation, we consider a simple approach. For example, the workload generated by an invocation j of PS at RP i is given in MIs by: $p_j * y_j * PSRL_j$ where p_j is the number of processors required by the PS, y_j is the processor speed in MIPS and $PSRL_j$ is PS request length.

Table 1. Description of the parameters Used in the Experiments

Parameter Description	Acronym	Exp. 1	Exp. 2	Exp. 3	Exp. 4
Number of Providers of Workflow Services		10	20	50	50
Number of Providers of Processing Basic Services		10	20	50	50
Number of Providers of Storage Basic Services		10	20	50	50
Number of Providers of Processing Resource Bundles		10	20	50	20
Number of Providers of Storage Resource Bundles		10	20	50	20
Number of Service Instances Per WFS Provider	SI		40		
Number of Resource Bundles Per Resource Provider	RU		1		
Number of Requests to Workflow Service	WSR		1000		
Time between arrivals of WFS requests	TBWS		0-120s		
WFS Request Length	WSRL		30-60s		
PS Request Length	PSRL		WSRL/2		
SS Request Length	SSRL		WSRL/2		
Input File Size	INSIZE		30-50KB		
Output File Size	OUTSIZE		100-200KB		

Table 1 summarises the experiments performed and the values used for the simulation of the service application in GridSim using the Catallaxy economic model and the presented service framework. The parameters TBWS, WSRL, INSIZE and OUTSIZE use uniform distributions. In addition, we consider that the BSPs are able to provide and negotiate for one BS at a time.

5.2   Experimental Results

Fig. 4(a) shows the allocation rate of workflow service requests in the different experiments. The allocation rate is above 96% in all experiments. However, in Experiment 3 the allocation rate is lower than in Experiment 4, even though more resource providers are available. The reason for such behaviour is that a provider reserves its services or resources when it receives a CFP. Once an announcement is sent by the provider who initiated the negotiation, the providers that have not been selected release their services or resources. As the number of providers increase, more messages are sent, the negotiations take more time and the resources are kept reserved for a longer time. In Experiment 4, we reduce the number of resource providers and the allocation rate increases.

We then evaluate the impact of the negotiations on the service provisioning process. The experiments measure the amount of time spent on negotiation for a BS. Fig. 4(b) shows the time spent in different scenarios. We observe that the time spent is highly dependent on the initial timeout during which the negotiator waits for proposals, which in this case is 30 seconds (15 seconds in negotiation for the BS and 15 seconds in negotiation for the resource). We thus omit this 30 second interval from the results presented in the figure. In the scenarios evaluated, we consider that users and service providers are in different networks connected through a network link with a bandwidth of 1Mbps while service providers and resource providers are connected through another network link with a bandwidth of 1Mbps. Both links present a latency of 50 milliseconds, which we consider to be representative of the latency in many wide area networks. The time required to send proposals and to bargain to achieve the final price is generally smaller than 10 seconds. The initial timeout can be reduced if the initial negotiator knows how many providers have been contacted and how many

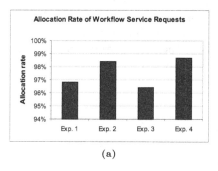

(a)

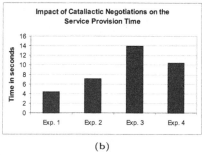

(b)

Fig. 4. (a)Allocation rate of WFS requests. (b)Time taken for a BS negotiation.

messages should be received. However, we envision a scenario in which a P2P network is used to broadcast calls for proposals and the negotiator does not know exactly how many providers will receive the proposals and send a reply.

6   Conclusion and Future Work

This paper describes a model for the simulation of service-oriented Grid applications to allow the decoupling of service negotiation and resource management into two distinct layers. By decoupling these, it is possible to model and evaluate different strategies for both service provisioning and resource allocation. The model also enables the simulation and evaluation of policies for negotiation of SLAs for service usage and the evaluation of centralised and decentralised economic models. We present experimental results that demonstrate the use of the framework for modelling and evaluation of a decentralised economic bargaining mechanism, the Catallaxy, for service and resource negotiation.

For future work, we would like to evaluate the suitability of the framework for modelling large-scale scenarios and improve the acquisition policies to support advance reservation and co-allocation of Grid resources. In addition, we would like to evaluate the economic models considering dynamic environments with varying failure probabilities for resources. We will consider acquiring data from existing Grid test beds for determining the failure probability of Grid resources and include these in the Grid simulator.

In addition, we would like to incorporate models for what can be called elastic containers or elastic VMs. In these models, the allocation policy of a utility data centre, for instance, may decide to expand the amount of memory, storage and CPU of VMs in an AE according to the service workload. We would like to incorporate these VM models and enable the changes in the configurations of VMs on the fly. These features can enable the evaluation of varying provisioning policies.

Acknowledgments

We thank Chee Shin Yeo, Krishna Nadiminti, James Broberg and Al-Mukaddim Khan Pathan from the University of Melbourne for their assistance in improving the quality of this paper and for sharing their thoughts on the topic. This work is supported by the European Union, DEST and ARC Project grants. Marcos' PhD research is partially supported by National ICT Australia (NICTA).

References

1. Foster, I.: Service-oriented science. science 308(5723), 814–817 (2005)
2. Kephart, J.O., Chess, D.M.: The vision of autonomic computing. Computer 36(1), 41–50 (2003)

3. Almeida, J., Almeida, V., Ardagna, D., Francalanci, C., Trubian, M.: Resource management in the autonomic service-oriented architecture. In: 3rd IEEE International Conference on Autonomic Computing (ICAC 2006), Dublin, Ireland, pp. 84–92 (2006)
4. Bennani, M.N., Menascé, D.A.: Resource allocation for autonomic data centers using analytic performance models. In: 2nd IEEE International Conference on Autonomic Computing (ICAC 2005), Seattle, Washington, pp. 229–240 (2005)
5. Walsh, W.E., Tesauro, G., Kephart, J.O., Das, R.: Utility functions in autonomic systems. In: International Conference on Autonomic Computing (ICAC 2004), New York, pp. 70–77 (2004)
6. Barham, P., Dragovic, B., Fraser, K., Hand, S., Harris, T., Ho, A., Neugebauer, R., Pratt, I., Warfield, A.: Xen and the art of virtualization. In: 19th ACM Symposium on Operating Systems Principles (SOSP '03), pp. 164–177. ACM Press, New York (2003)
7. Low, C., Byde, A.: Market-based approaches to utility computing. Technical Report HPL-2006-23, Internet Systems and Storage Laboratory, Hewlett Packard Laboratories Bristol (2006)
8. Bell, W.H., Cameron, D.G., Capozza, L., Millar, A.P., Stockinger, K., Zini, F.: Simulation of dynamic grid replication strategies in optorsim. In: Parashar, M. (ed.) GRID 2002. LNCS, vol. 2536, pp. 46–57. Springer, Heidelberg (2002)
9. Casanova, H.: Simgrid: A toolkit for the simulation of application scheduling. In: 1st IEEE/ACM International Symposium on Cluster Computing and the Grid (CCGrid 2001), Brisbane, Australia, pp. 430–437 (2001)
10. Song, H., Liu, X., Jakobsen, D., Bhagwan, R., Zhang, X., Taura, K., Chien, A.: The microgrid: a scientific tool for modeling computational grids. In: Reich, S., Anderson, K.M. (eds.) Open Hypermedia Systems and Structural Computing. LNCS, vol. 1903, p. 53. Springer, Heidelberg (2000)
11. Buyya, R., Murshed, M.: Gridsim: A toolkit for the modeling and simulation of distributed resource management and scheduling for grid computing. Concurrency and Computation: Practice and Experience (CPE) 14(13-15), 1175–1220 (2002)
12. Grit, L., Inwin, D., Yumerefendi, A., Chase, J.: Virtual machine hosting for networked clusters: Building the foundations for 'autonomic' orchestration. In: 1st International Workshop on Virtualization Technology in Distributed Computing (VTDC 2006), Tampa, Florida (2006)
13. Urgaonkar, B., Roscoe, P.S.T.: Resource overbooking and application profiling in shared hosting platforms. In: 5th Symposium on Operating Systems Design and Implementation, Boston, Massachusetts, pp. 239–254 (2002)
14. Hayek, F.A.V.: The Collected Works of F.A. Hayek. University of Chicago Press, Chicago (1989)
15. Eymann, T., Ardaiz, O., Catalano, M., Chacin, P., Chao, I., Freitag, F., Gallegati, M., Giulioni, G., Joita, L., Navarro, L., Neumann, D.G., Rana, O., Reinicke, M., Schiaffino, R.C., Schnizler, B., Streitberger, W., Veit, D., Zini, F.: Catallaxy-based grid markets. International Journal on Multiagent and Grid Systems, Special Issue on Smart Grid Technologies & Market Models 1(4), 297–307 (2005)
16. Reinicke, M., Streitberger, W., Eymann, T.: Scalability analysis of matchmakers in self-optimizing computing networks. Journal of Autonomic and Trusted Computing (JoATC) (2005)

Managing a Peer-to-Peer Backup System: Does Imposed Fairness Socially Outperform a Revenue-Driven Monopoly?

László Toka and Patrick Maillé

GET/ENST Bretagne
2, rue de la Châtaigneraie CS 17607
35576 Cesson Sévigné Cedex, France
{laszlo.toka, patrick.maille}@enst-bretagne.fr

Abstract. We study a peer-to-peer backup system, where users offer some of their storage space to provide service for the others. The economic model for such a system is different from the ones applicable to peer-to-peer file sharing systems, since the storage capacity is a private good here. We study two mechanisms aimed at incentivizing users to offer some of their capacity: a price-based scheme (here a revenue-driven monopoly) and a more classical symmetric scheme (imposing users to contribute to the service at least as much as use it). We compare the outcomes of such mechanisms to the socially optimal situation that could be attained if users were not selfish, and show that depending on user heterogeneity, a revenue maximizing monopoly can be a worse or a better (in terms of social welfare) way to manage the system than a symmetric scheme.

Keywords: Peer-to-peer networks, economics, incentives, pricing.

1 Introduction

With the convergence of fixed and mobile telecommunication systems, all kinds of digital documents (e.g. videos and audio files, e-mails) are likely to be accessed by different types of devices (mobile phone, personal computer, mp3 player). The storing of all those then documents raises several questions: should there be only one storing location? If so, what happens in case of a crash? If not, how to update documents between several locations? Will it be simple to transfer a document from a storing location to a given device?

In this paper, we suggest that those problems be addressed via a distributed storing system working in a peer-to-peer (P2P) way: using a P2P network infrastructure, a (ciphered) copy of each user's data is stored into the hard drives of other participants in the network. As in peer-to-peer (P2P) file sharing networks, each participant is consequently at the same time a service user and a service provider. Such a service presents a lot of advantages in terms of reliability (data replication within the system provides a protection against failures) and ease of access (each user can access his data from any device connected to the network).

D.J. Veit and J. Altmann (Eds.): GECON 2007, LNCS 4685, pp. 150–163, 2007.

A peer-to-peer backup system has already been proposed and studied in [1], that introduces pStore, a secure distributed backup system based on an adaptive P2P network. pStore exploits unused personal hard drive space attached to the Internet to provide the distributed redundancy needed for reliable and effective data backup. Moreover, support for file encryption, versioning, and secure sharing is provided. Nevertheless, no study on how users would react to such a system is carried out. This paper intents to investigate that particular issue.

Indeed, it seems reasonable to us to assume that each user is selfish, i.e. is only sensitive to the quality of service she experiences, regardless of the effects of her actions on the other users. The framework of *Game Theory* [2] is therefore particularly well-suited to study that kind of interaction among agents: the situation is then studied as a non-cooperative game played among users, where a user strategy is the amount of memory capacity offered to provide service to the other users and the amount of data she stores into the system. Notice that as for other peer-to-peer (P2P) applications, a user valuation for the service depends on the "generosity" of the other users: each user benefits from the others' shared capacity. However, there is no direct incentive to offer one's own capacity to the others, and users are then incentivized to free-ride [3], i.e. benefit from the service without contributing to it: if the sharing efforts do not get some kind of proof of appreciation, nobody has interest to cooperate and the service cannot exist. Therefore it is necessary that some incentive mechanisms be properly designed for the service to actually exist and be valuable for users.

For P2P file sharing systems, there is growing evidence of that need for incentives. For instance, one study of the Gnutella file sharing system showed that almost 70% of the peers only consume resources but do not provide any files [3]. The problem of incentivizing users to contribute in such systems has been the subject of extensive research [4,5,6,7,8].

On the other hand, the existing literature on P2P backup systems mainly focuses on security, reliability and technical feasibility issues [9,10], whereas the incentive aspect received little attention. Notice also that the economic models developed for P2P file sharing systems do not apply to P2P backup services: in file sharing systems, when a peer provides some files to the community, she contributes to the whole system in terms of accessibility. This means that the resource is not dedicated to a certain number of users, but is offered to all the rest of the peers, and in that sense the information stored in the P2P network is a public good. On the contrary, a P2P backup system operates on non-divisible resources, i.e. a certain disk space belongs to one given user (for the time being) and no other peer can access it. The storing resource available on the network is then a private good, and it cannot be managed the same way as a public good from an economic point of view.

The existing models for P2P backup services focus on solutions that do not require financial transactions. Therefore the counter payment for a given service is usually the service in question as well. This approach finally leads to a symmetric scheme where every peer should contribute to the system in terms of service at least as much as she benefits from others [11,12,13].

The incentive part of the scheme proposed in [10] relies on the use of a "probation" period, during which a peer must prove herself reliable before benefiting from the system. A very strict policy based on quotas is suggested in [14]: each peer (identified by her IP address) cannot insert more than a given amount of data into the system. Likewise, the distributed accounting infrastructure proposed in [15] proceeds by simple exclusion of non-cooperating peers from the system, that are detected via an audit.

In this paper, we investigate more flexible solutions, that could still provide peers with incentives to contribute to the system. We focus on the performance of incentive schemes. We propose to study and compare two types of incentive mechanisms that have been suggested in the literature in other - but linked - contexts, like file sharing systems, connection sharing systems, and ad hoc networks: some of those schemes rely on monetary incentives, and some others are based on service degradation for users who do not contribute enough to the service. A particular instance of each type of scheme is considered, and their effects on the overall social welfare are weighted for the particular context of the P2P backup service.

The paper is organized as follows. The model we consider for user preferences is depicted in Section 2, where we also study the maximum reachable value of social welfare yielded by the service. A strict symmetry-based scheme is studied in Section 3, and schemes implementing pricing are investigated in Section 4. The performance of those schemes in terms of social welfare are compared in Section 5. Section 6 presents our conclusions and directions for future work.

2   Model

In this section we describe the model we consider in this paper. We first introduce utility functions that represent user preferences and the decision variables that constitute user strategies. Then we consider the "ideal" situation where users would not behave selfishly and act so as to maximize the total system performance (social welfare). That ideal situation will be used in the next sections as a reference to study incentive schemes.

Note that in this paper, we say that the amount of storage space that is necessary to safely store some data in the system equals the size of those data. This is done without loss of generality: assume that the system introduces a redundancy factor r to improve the data availability on the system, then this is equivalent to replacing C_i^s by C_i^s/r in the user cost functions, or equivalently to replacing C_i^o by rC_i^o in the user valuation function (remark that prices have a different interpretation depending on that choice: they are per unit of physical capacity in the former case, and per unit of "sufficiently available" capacity - taking into account the redundancy - in the latter).

2.1   User Utility Function

We provide here a model for user preferences. The user set is denoted by $\mathcal{I}$, and the perceived utility for a user $i \in \mathcal{I}$ offering capacity C_i^o, storing an amount C_i^s of data in the system and paying a total charge π_i should be a decreasing function of C_i^o and an increasing function of C_i^s. We suggest to use a separable additive function of the utility perceived by a user i, as described in the following definition.

Definition 1. *The utility U_i of a user $i \in \mathcal{I}$ is of the form*

$$U_i\left(C_i^o, C_i^s, \pi_i\right) = V_i(C_i^s) - P_i(C_i^o) - \pi_i, \tag{1}$$

where

- $V_i(C_i^s)$ *is the valuation of user i, i.e. the price she is willing to pay to store an amount C_i^s of data in the system. In this paper we will assume that $V_i(\cdot)$ is positive, continuously differentiable, increasing and concave in its argument, and that $V_i(0) = 0$ (no service yields no value).*
- $P_i(C_i^o)$ *is the opportunity cost of user i for offering capacity C_i^o to the system, i.e. it is the price that she is willing to be paid to devote C_i^o of her disk space to provide service. We assume that $P_i(\cdot)$ is positive, continuously differentiable, increasing and strictly convex, and that $P_i(0) = 0$ (no contribution brings no cost).*

From the valuation and cost functions, we can be derive (by differentiation and taking the inverse functions) two other functions.

Definition 2. *For a user $i \in \mathcal{I}$, we call* demand function *(resp.* supply function*) the function $d_i(\cdot)$ (resp. $s_i(\cdot)$) such that for all $p \in \mathbb{R}_+$,*

$$d_i(p) := \begin{cases} (V_i')^{-1}(p) & \text{if } p \le V_i'(0), \\ 0 & \text{otherwise.} \end{cases} \tag{2}$$

$$s_i(p) := \begin{cases} (P_i')^{-1}(p) & \text{if } p < \lim_{q \to +\infty} P_i'(q), \\ +\infty & \text{otherwise,} \end{cases} \tag{3}$$

where f' stands for the derivative function of function f.

For a given $p \ge 0$, $d_i(p)$ (resp. $s_i(p)$) is the amount of storage capacity that user i would choose to use from (resp. to offer to) the others if she is charged (resp. paid) a unit price p for it.

Remark that, as intuitively expected, the demand (resp. supply) function is nonnegative and decreasing (resp. increasing) in the unit price.

To carry out a deeper analysis in the next sections, we will assume that the demand functions are of the same form for all users, and only differ through a multiplicative constant. Likewise, we make the same assumption regarding the supply functions.

Assumption A (Common form of supply and demand functions)
There exist a nonnegative and nonincreasing "common" demand function $d(\cdot)$, and a nonnegative and nondecreasing "common" supply function $s(\cdot)$ such that for all user $i \in \mathcal{I}$ there are positive real values a_i and b_i which satisfy

$$d_i = a_i \times d \qquad (4)$$
$$s_i = b_i \times s \qquad (5)$$

Moreover,

- $d(0) > 0$ *and* $s(0) = 0$
- $d(\cdot)$ *is strictly decreasing while it takes strictly positive values.*
- $s(\cdot)$ *is strictly increasing (eventually up to a point after which it is constant).*

Notice that the same kind of assumption (i.e. same form of utility functions for all users) is made in [16] in the framework of a P2P file sharing system, for user valuation functions.

Some of our results in the next sections are established for linear demand and supply functions d and v.

Assumption B (Affine demand and supply functions)
- *The common demand function d is affine. More precisely, there exists $\bar{p} > 0$ such that $d(p) = [\bar{p} - p]^+$, where for $y \in \mathbb{R}$, $y^+ = \max(0, y)$.*
- *The common supply function s is linear, i.e. $s(p) = p$.*

Under Assumptions A and B, the demand and supply functions of a user i express as follows:

$$d_i(p) = a_i[\bar{p} - p]^+, \qquad (6)$$
$$s_i(p) = b_i p. \qquad (7)$$

This corresponds to quadratic functions for the valuation and cost functions:

$$V_i(C_i^s) = \frac{1}{a_i}\left(-\frac{(C_i^s \wedge a_i\bar{p})^2}{2} + a_i\bar{p}\,(C_i^s \wedge a_i\bar{p})\right) \qquad (8)$$

$$P_i(C_i^o) = \frac{1}{b_i}\frac{C_i^{o2}}{2}, \qquad (9)$$

where $\wedge$ denotes the min. Finally, we will sometimes consider the following assumption in the case of a large number of users.

Assumption C. *Under Assumption A, the values a_i (resp. b_i) of all users $i \in \mathcal{I}$ are independent and identically distributed. Moreover, a_i and b_i are independent.*

2.2   Social Welfare

A user can choose her own strategy by varying her C_i^s and C_i^o parameters[1]. In this subsection we define social welfare, which will be used later as a performance measure to compare different incentive schemes.

Definition 3. *We call* social welfare *(or* welfare*) and denote by* W *the sum of the utilities of all agents in the system:*

$$W := \sum_i V_i(C_i^s) - P_i(C_i^o). \tag{10}$$

Notice that no prices appear in (10). This is because even if we consider a payment-based incentive scheme, we choose to include in social welfare all system agents, eventually including the entity that receives (or gives) payments. The utility of this entity would be its revenue, and all money it exchanges with the users would stay within the system and therefore does not influence social welfare.

Let us have a look at the "optimal" situation that the system can attain (in terms of social welfare maximization). The problem expresses

$$\max_{C_i^s, C_i^o} \left(\sum_i V_i(C_i^s) - P_i(C_i^o) \right), \tag{11}$$

subject to $C_i^o \geq 0$, $C_i^s \geq 0$ for $\forall i$, and

$$\sum_i C_i^o \geq \sum_i C_i^s. \tag{12}$$

This is a classical convex optimization problem, that can be solved by the Lagrangian method.

- The first order conditions imply that for all $i \in \mathcal{I}$, $P_i'(C_i^o) = p$ and $V_i'(C_i^s) = p$, where $p \geq 0$ is the Lagrange multiplier relative to the feasibility constraint (12).
- The complementary slackness condition writes $\min \left[p, \sum_i (C_i^o - C_i^s) \right] = 0$.

Moreover, p must be strictly positive: otherwise the first order conditions give $C_i^o = 0$ and $C_i^s > 0$ for all i, violating the feasibility constraint (12). We therefore obtain

$$C_i^s = d_i(p^*), \qquad C_i^o = s_i(p^*), \tag{13}$$

---

[1] Since staying online induces a disutility for a user without direct counterpart but improves the quality of the service offered to the others, incentives are needed to honor peers that are online almost all the time. The definition of such incentives is ongoing work, and is out of the scope of this paper. We will not consider it here, assuming that users stay online as much as they can, without trying to minimize the associated costs.

where p^* is the (unique) solution of

$$\sum_i s_i(p^*) - d_i(p^*) = 0,$$ (14)

and the optimal value of the social welfare is then

$$W^* = \sum_{i \in \mathcal{I}} V_i(d_i(p^*)) - P_i(s_i(p^*)).$$ (15)

Figure 1 (displayed in Section 4) gives a graphical interpretation of the maximum social welfare that can be attained by the system. Notice that the Lagrange multiplier can be interpreted as a unit price: if users buy the resource at unit price p^*, and sell their available disk capacity at the same unit price, then the selfish user decisions drive the system to the welfare maximizing solution.

The following result considers our particular assumptions.

Proposition 1. *Under Assumptions A and B, the maximal value W^* of social welfare is*

$$W^* = \frac{1}{2} \bar{p}^2 \frac{\sum_i a_i \sum_i b_i}{\sum_i (a_i + b_i)}.$$ (16)

Proof. From (14), we get the social welfare at the ideally fine-tuned unit price:

$$p^* = \bar{p} \frac{\sum_i a_i}{\sum_i a_i + b_i}.$$ (17)

Therefore we have $C_i^o = b_i p^* = \bar{p} \frac{b_i \sum_{j \in \mathcal{I}} a_j}{\sum_{j \in \mathcal{I}} a_i + b_i}$ and $C_i^s = \bar{p} \frac{a_i \sum_{j \in \mathcal{I}} b_j}{\sum_{j \in \mathcal{I}} a_j + b_j}$, which gives, after some simplifications,

$$W^* = \frac{1}{2} \bar{p}^2 \frac{\sum_i a_i \sum_i b_i}{\sum_i (a_i + b_i)},$$ (18)

and establishes the proposition.

We therefore have a characterization of the optimal solution. However, as pointed out in the introduction, user selfishness does not lead to this optimal situation when users are not incentivized to offer service to the others. Actually, the unique Nash equilibrium of the game without incentives corresponds to the situation where $C_i^o = 0$ for all i, and the associated social welfare is 0.

In the rest of the paper, we investigate two kinds of incentive schemes, and study their performance in terms of social welfare. We first consider mechanisms without pricing, that simply impose users to provide at least as much memory space as the amount they intend to use (most of the existing related works support this kind of fairness providing approach). Then we turn to payment-based incentive mechanisms where users have to pay for using the service and are paid if they contribute. We finally compare the outcomes of those schemes in terms of social welfare, for some particular types of valuation and cost functions. Since for the two schemes under study the optimal situation cannot be reached in general, we measure the loss of welfare of those schemes with respect to the maximum value.

3   Performance of Schemes Imposing Symmetry

In this section, we follow the ideas suggested in the literature for schemes without pricing. As evoked in the introduction, the principle of those schemes is that users are invited to contribute to, at least as much as they take from, the others. Each user i then chooses C_i^o and C_i^s so as to maximize $V_i(C_i^s) - P_i(C_i^o)$, subject to $C_i^o \geq C_i^s$. As $P_i(\cdot)$ is increasing in C_i^s, no user has an interest to choose a strategy with $C_i^o > C_i^s$. Therefore a user will necessarily choose $C_i^o = C_i^s$. User i maximizes her utility[2] at the point $C_i^s = C_i^o = C_i^*$ where

$$V_i'(C_i^*) - P_i'(C_i^*) = 0. \tag{19}$$

Under our specific assumptions on demand and supply functions, the value of social welfare for such a scheme can be derived:

Proposition 2. *Under Assumptions A and B, the ratio of the social welfare for the symmetric scheme W_{sym} to the maximum social welfare W^* is*

$$\frac{W_{sym}}{W^*} = \left(\frac{1}{\sum_i a_i} + \frac{1}{\sum_i b_i} \right) \sum_i \left[\frac{1}{\frac{1}{a_i} + \frac{1}{b_i}} \right]. \tag{20}$$

Moreover, under Assumption and C, this ratio converges as the number of users tends to infinity

$$\frac{W_{sym}}{W^*} \xrightarrow[|\mathcal{I}| \to \infty]{} \left(\frac{1}{\mathbb{E}[a]} + \frac{1}{\mathbb{E}[b]} \right) \mathbb{E}\left[\frac{1}{\frac{1}{a} + \frac{1}{b}} \right] \tag{21}$$

Proof. We straightforwardly obtain that for all i, $C_i^o = C_i^s = \bar{p}\frac{a_i b_i}{a_i + b_i}$. The corresponding social welfare is then $W_{sym} = \frac{1}{2}\bar{p}^2 \sum_{i \in \mathcal{I}} \frac{a_i b_i}{a_i + b_i}$, and (20) then comes from Proposition 1. The law of large numbers gives (21).

As $f : x, y \mapsto \frac{1}{\frac{1}{x} + \frac{1}{y}}$ is strictly concave, Jensen's inequality implies that $W_{opt} \geq W_{sym}$, and that equality stands if and only if a, b are deterministic, i.e. identical for every user.

4   Performance of Pricing Mechanisms

In this section, we study the influence of introducing a specific pricing scheme for incentivizing users to offer storage capacity, and preventing them from using more capacity than what is available. We consider a simple mechanism: contributors are paid p^o per unit of storage capacity they offer to the system, and service users are charged a unit price p^s when they store their data onto the

---

[2] Actually, the utility maximization problem for a user is a convex problem, that has the same form as the social welfare maximization problem studied in subsection 2.2, except that there are only two decisions variables (C_i^o and C_i^s) here.

system. Such a mechanism offers users the choice to act as a pure consumer, as a pure service provider, or to both contribute to and benefit from the service. Remark that we will not try here to avoid the presence of a central authority or clearance service: as the model aims to give hints for a commercial application, it is reasonnable to consider the existence of such an entity.

The amount that user i will be charged (this amount can be negative, in which case the user gets paid) is consequently

$$\pi_i = p^s C_i^s - p^o C_i^o. \tag{22}$$

We analyze the model as a full information game, i.e. we assume that the entity that operates the service (the operator) has perfect knowledge of the users and their valuation and cost functions. Therefore, knowing that users will act so as to maximize their utility, it can predict user reactions, and drive the outcome of the game to the most profitable situation for itself. In this sense, the operator acts as the leader of a Stackelberg (or leader-follower) game [2]. We investigate two possibilities: either the coordinator aims at maximizing the user surplus, or it is a revenue-driven monopoly that chooses prices so as to maximize its revenue. In both cases, the feasibility constraint (12) must be satisfied.

Welfare-Maximizing Operator. From our study in subsection 2.2, the theoretically highest level of social welfare can be attained by a payment based scheme. In fact it is reached when the selling and buying prices are the same and equal p^* (see (14)). In that case the operator monetary surplus is null: the operator has no income at all and just acts as a coordinator that redistributes money among users.

Profit-Oriented Monopoly. In this subsection, we assume that the monopoly strives to extract the maximum profit out of the business. The operator therefore faces the following maximization problem.

$$\max_{p^s, p^o} \left(p^s \sum_i d_i(p^s) - p^o \sum_i s_i(p^o) \right), \tag{23}$$

subject to $p^s \geq 0$, $p^o \geq 0$ and the constraint (12) that writes $\sum_i s_i(p^o) \geq \sum_i d_i(p^s)$.

This problem is hard to solve for general utility functions, and even under Assumption A since it is not a convex problem. We therefore consider the case where Assumption B holds.

Proposition 3. *Under Assumptions A and B the performance ratio of the social welfare for a profit-maximizing monopoly W_{mon} to the maximum social welfare W^* is*

$$\frac{W_{mon}}{W^*} = \frac{3}{4}. \tag{24}$$

Proof. Under Assumptions A and B, the profit maximization problem becomes a convex problem that can be solved using the Lagrange method for example. However here a simple graphical argument is enough to conclude.

Figure 1 plots two curves: the total demand $D = \sum_i d_i$ and total supply $S = \sum_i s_i$ as functions of the unit price p. First remark that p^o and p^s must be chosen such that $S(p^o) = D(p^s)$: otherwise it is always possible for the operator to decrease p^o (if $S(p^o) > D(p^s)$) or increase p^s (if $S(p^o) < D(p^s)$) to strictly improve its revenue. The operator revenue with such prices is then the area of the rectangle displayed in Figure 1, embedded within a triangle whose area is the maximum value of social welfare. The area of the rectangle is maximum when $S(p^o) = D(p^s) = Q^*/2$ with $Q^* = Dp^*$ and p^* is given in (14). In that case the operator's profit is $W^*/2$, and the total social welfare is $3W^*/4$.

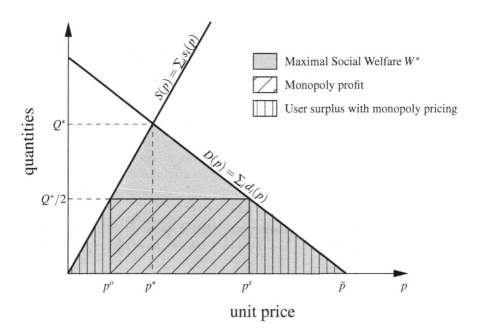

Fig. 1. Maximal social welfare, and optimal choices for a revenue-driven monopoly under Assumptions A and B

5   Does Imposed Symmetry Outperform Monopoly Pricing?

In this section, we compare the two incentive schemes introduced so far in terms of the underlying social welfare at equilibrium. Our point here is not to give necessary and sufficient conditions for one of the two mechanism to provide a larger social welfare than the other for general valuation and cost functions, since the problem becomes much more difficult when Assumptions A and B are relaxed (in particular some optimization problems like revenue maximization are nonconvex and may exhibit local optima). We rather concentrate on our simplifying assumptions and use some examples to highlight situations where one scheme or the other can be better for the overall system.

The following result is a direct consequence of Propositions 2 and 3:

Proposition 4. *Under Assumptions A, B, and C, and for a large number of users, symmetric schemes outperform monopoly pricing if and only if*

$$\left(\frac{1}{\mathbb{E}[a]} + \frac{1}{\mathbb{E}[b]} \right) \mathbb{E}\left[\frac{1}{\frac{1}{a} + \frac{1}{b}} \right] \geq \frac{3}{4}. \tag{25}$$

Proposition 4 highlights in particular the fact that if the population is homogeneous (i.e., a and b are Dirac distributions) then it is better to implement a scheme based on symmetry, since the maximal social welfare can be attained (from Jensen's equality case). On the contrary, user heterogeneity in terms of a and b will make the left-hand side of (25) decrease. If heterogeneity is too important then the left-hand side of (25) may take values below $3/4$, which implies that the system (users+coordinator) is better off being driven by a revenue-maximizing monopoly.

At this point of the analysis the distribution of a and b turns out to be the main characteristics of the game. Indeed, these parameters characterize the profile of each user, i.e. her utility and cost functions, and her associated demand and supply functions $(d_i(p), s_i(p))$. In the following we consider two simple examples of distributions (e.g., uniform and exponential) for a and b to illustrate Proposition 4.

Uniform Distribution. We assume here that a (resp. b) is uniformly distributed over $[0, a_{\max}]$ (resp. $[0, b_{\max}]$). In that case we have $\mathbb{E}[a] = \frac{a_{\max}}{2}$, $\mathbb{E}[b] = \frac{b_{\max}}{2}$, and

$$\mathbb{E}\left[\frac{1}{\frac{1}{a} + \frac{1}{b}} \right] = \frac{1}{3}\left(a_{\max} + b_{\max} - \frac{a_{\max}^2}{b_{\max}} \ln(1 + \frac{b_{\max}}{a_{\max}}) - \frac{b_{\max}^2}{a_{\max}} \ln(1 + \frac{a_{\max}}{b_{\max}}) \right). \tag{26}$$

The left-hand side of (25) and the plane $z = 3/4$ are displayed on Figure 2 *(left)*. We observe that inequality (25) always holds, thus it is always better for the system to impose symmetry than to introduce a profit-maximizing monopoly.

Exponential Distribution. We now consider the case where a (resp. b) follows an exponential distributions with parameter μ_a (resp. μ_b). Therefore $\mathbb{E}[a] = \frac{1}{\mu_a}$, $\mathbb{E}[b] = \frac{1}{\mu_b}$, and after some calculation we obtain

$$\mathbb{E}\left[\frac{1}{\frac{1}{a} + \frac{1}{b}} \right] = \begin{cases} \frac{1}{3\mu_a} & \text{if } \mu_a = \mu_b, \\ \frac{\mu_a^2 - \mu_b^2 - 2\mu_a\mu_b \ln(\frac{\mu_a}{\mu_b})}{(\mu_a - \mu_b)^3} & \text{otherwise.} \end{cases} \tag{27}$$

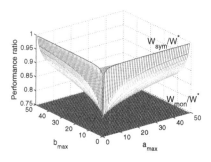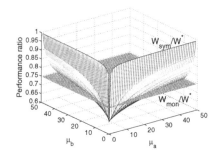

Fig. 2. Social welfare performance ratio of monopoly pricing *(plane surfaces)* and symmetric scheme *(curved surfaces)* as given by Propositions 2 and 3, for uniform *(left)* and exponential *(right)* distributions of a and b

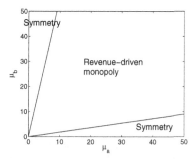

Fig. 3. Best scheme (in the sense of social welfare) for exponential distributions of a and b (μ_a and μ_b are the parameters of the exponential laws)

We again compare both terms of (25) on Figure 2 *(right)*. This time, there is no scheme that always outperforms the other: depending on how much the two variables differ, the monopoly can provide higher welfare than a symmetric scheme. More precisely, when μ_a and μ_b are sufficiently close, then a revenue-maximizing monopoly will drive the system to a situation where the social welfare is larger than what a symmetry-based scheme would have yielded. Figure 5 indicates the scheme that gives the best system welfare, depending on the values of μ_a and μ_b.

6   Conclusions and Future Work

In this paper, we have proposed an economic model for a peer-to-peer backup service. Assuming that users of such a service behave selfishly, we have justified the need for incentive schemes for the system to effectively exist. We have studied and compared two kinds of incentive schemes, namely a symmetry-based scheme (users should contribute to the service as much as they use it) and a pricing-based scheme (introduction of a monopoly that fixes unit prices for buying and selling resource). Under some simplifying assumptions, we have highlighted conditions for one scheme to outperform the other in terms of social welfare. Some examples have shown that user heterogeneity plays a crucial role on the better-suited scheme. Basically, it seems that more user heterogeneity would justify the use of (even profit-driven) pricing.

There remains a lot of work to be done on this subject. First, we have made quite restrictive assumptions on the form of the utility functions to derive some results (we often used quadratic valuation and cost functions). It would be interesting to obtain some results for more general cases. Moreover, we only considered that users had two decision variables, namely the quantity of data they store into the system and the amount of storing capacity they offer. We are currently working toward an extension of the model where users can also choose the proportion of time they are online (a larger availability improves the service offered to the others, but increases the perceived cost of the user), that also needs an incentive scheme.

Acknowledgment. This work has been partially funded by the GET project DisPairSe.

References

1. Batten, C., Barr, K., Saraf, A., Treptin, S.: pStore: A secure peer-to-peer backup system. Technical Report MIT-LCS-TM-632, MIT Laboratory for Computer Science (2001)
2. Fudenberg, D., Tirole, J.: Game Theory. MIT Press, Cambridge, Massachusetts (1991)
3. Adar, E., Huberman, B.: Free riding on gnutella. Tech. rep. Xerox parc (2000)
4. Anagnostakis, K., Greenwald, M.: Exchange-based incentive mechanisms for peer-to-peer file sharing. In: Proc. of 24th International Conference on Distributed Computing Systems (ICDCS'04), Tokyo, Japan (2004)
5. Cohen, B.: Incentives build robustness in bittorrent. In: Proc. of 1st Workshop on Economics of Peer-to-Peer Systems (P2PECON'03), Berkeley, CA (2003)
6. Golle, P., Leyton-Brown, K., Mironov, I., Lillibridge, M.: Incentives for sharing in peer-to-peer networks. In: Proc. of 3rd ACM conference on Electronic Commerce (EC'01), Tampa, Florida, pp. 264–267 (2001)
7. Lai, K., Feldman, M., Stoica, I., Chuang, J.: Incentives for cooperation in peer-to-peer networks. In: Proc. of Workshop on Economics of P2P Systems (2003)
8. Vishnumurthy, V., Chandrakumar, S., Sirer, E.: Karma: A secure economic framework for peer-to-peer resource sharing. In: Proc. of 1st Workshop on Economics of Peer-to-Peer Systems (P2PECON'03), Berkeley, CA (2003)
9. Druschel, P., Rowstron, A.: PAST: A large-scale, persistent peer-to-peer storage utility. In: HotOS VIII, Schloss Elmau, Germany, pp. 75–80 (2001)
10. Lillibridge, M., Elnikety, S., Birrell, A., Burrows, M., Isard, M.: A cooperative internet backup scheme. In: Proc. of 1st Workshop on Economics of Peer-to-Peer Systems (P2PECON'03), Berkeley, CA (2003)

11. Cox, L., Noble, B.: Pastiche: Making backup cheap and easy. In: Proc. of Fifth USENIX Symposium on Operating Systems Design and Implementation, Boston, MA (2002)
12. Cox, L., Noble, B.: Samsara: Honor among thieves in peer-to-peer storage. In: Proc. of 19th ACM Symposium on Operating Systems Principles (SOSP'03), Bolton Landing, NY (2003)
13. Stefansson, B., Thodis, A., Ghodsi, A., Haridi, S.: MyriadStore. Technical Report T2006:09, Swedish Institute of Computer Science (2006)
14. Dabek, F., Kaashoek, M.F., Karger, D., Morris, R., Stoica, I.: Wide-area cooperative storage with CFS. In: Proc. of 18th ACM Symposium on Operating Systems Principles (SOSP'01), Chateau Lake Louise, Banff, Canada (2001)
15. Ngan, T., Nandi, A., Singh, A., Wallach, D., Druschel, P.: On designing incentives-compatible peer-to-peer systems. In: Proc. of 2nd International Workshop on Future Directions in Distributed Computing (FuDiCo II), Bertinoro, Italy (2004)
16. Courcoubetis, C., Weber, R.: Incentives for large peer-to-peer systems. IEEE JSAC 24(5), 1034–1050 (2006)

E-Business in ArguGRID

Francesca Toni

Imperial College London, Department of Computing
South Kensington Campus, London SW7 2AZ, UK
f.toni@imperial.ac.uk

Abstract. The ArguGRID project aims at supporting e-business applications, and in particular e-procurement, earth observation and business migration scenarios, by means of argumentative agent-based grid technologies. In this paper we outline the main features of the ARGUGRID envisaged system, intended to support more generally service selection and composition in distributed environments, including the grid and service-oriented architectures.

Keywords: e-business, service selection, service composition.

1 Introduction

The ArguGRID project aims at developing a grid-based platform populated by rational decision-making agents that are associated with service requestors/providers and users. Within agents, argumentation [2,3,4] is used to support decision making, taking into account (and despite) the often conflicting information that these agents have, as well as the preferences of users, service requestors and providers. Argumentation is also intended to support the negotiation between agents [8,10], on behalf of service requestors/providers/users. This negotiation takes place within dynamically formed virtual organisations. The agreed combination of services, amongst the agents, can be seen as a complex service within a service-centric architecture [1]. We intend to validate this overall approach by way of industrial application scenarios.

We have chosen to focus on e-business applications as we believe that they will benefit from a grid-based realisation, while at the same time illustrating and making use of the "semantic" techniques envisaged by ARGUGRID. These applications are intended to provide context for the project and to guide the development of formal models, their implementation, and subsequent experiments. Concretely, the chosen e-business scenarios are:

1. e-Procurement applications and e-Marketplaces,
2. e-Business for Earth Observation applications, and
3. the problem of business planning and outsourcing to new countries

The eProcurement and Earth Observation scenarios and use cases are the outcome of and build upon the extensive field experience of the respective industrial partners of the consortium (cosmoONE Hellas Market-site S.A. Greece and GMV S.A, Spain,

D.J. Veit and J. Altmann (Eds.): GECON 2007, LNCS 4685, pp. 164–169, 2007.

respectively) and the concrete use cases are realistic ones. Instead, the business planning and outsourcing scenario is the outcome of an analysis of academic non-practitioners (Asian Institute of Technology, Thailand), albeit the result of extensive market and field analysis.

In this paper we outline the ArguGRID approach and the main features of our chosen e-business scenarios.

2 ArguGRID: An Overview

ArguGRID aims to:

- develop argumentation-based foundations for the GRID, populated by rational decision-making agents within virtual organisations.
- incorporate argumentation models into service-centric architecture.
- develop underlying platform using peer-to-peer computing and overlay networks.
- validate the ArguGRID approach by way of industrial application scenarios.

This perspective is pictured in Fig. 1. The top layer is about building applications, focusing on e-business scenarios. The middle layer concerns the development of individual agents as well as methodologies for dynamically assembling agents into virtual organizations. These agents are responsible for the negotiation of contracts regulating the interactions amongst the agents to support the applications. We envisage that the agents will be able to resolve any disputes concerning the execution of the contracts, and that they may rely upon reputation measures during the operation of virtual organizations, as well as information about competence of the various agents [7]. Agents and virtual organizations "sit" on top of a service-centric architecture and the grid.

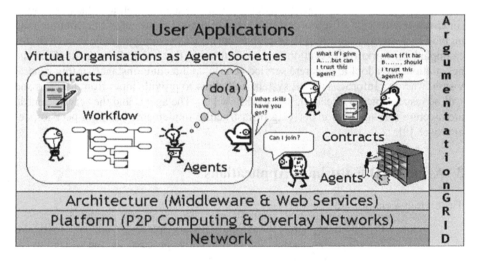

Fig. 1. ArguGRID perspective

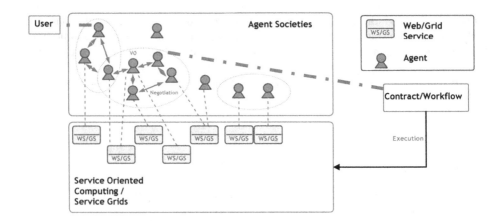

Fig. 2. The ArguGRID envisaged system

The envisaged system to support this vision is schematically described in Fig. 2.

Each service requestor/provider and each user is associated with one or more agents. Agents use argumentation for negotiating on behalf of service requestors/ providers/users. Users can provide input to agents, in terms of their objectives (what they expect to achieve from the service composition performed by the agents) and preferences (either for the specific objectives, or, more generally, as a generic profile of the user).

Agents negotiate with one another by using argumentation to support their decision making and communication processes. Negotiation takes place within dynamically created and maintained virtual organizations, envisaged as societies of agents whereby interaction is regulated by social norms and/or protocols. The outcome of negotiation results into a contract, understood, at the agent level, as a task allocation (in terms of provision of resources/services) to agents. In particular, this contract may include a workflow description [6], that needs to be appropriately executed, for example by a workflow execution engine. In the case that a workflow results from the negotiation, we adopt a concrete service-centric architecture, instance of this general vision, whereby InforSense KDE system will allow to provide input from the user and will be responsible for executing the workflow [1]. The agents and the service centric architecture rely upon an underlying infrastructure implemented using a peer-to-peer approach [9].

3 ArguGRID E-Business Applications

The chosen scenarios are

1. e-Procurement applications and e-Marketplaces, and in particular supporting the user's decision-making process in the selection of electronic auctions and the support of calls for proposals in e-Procurement

2.   e-Business for Earth Observation applications, and in particular satellite and sensor selection for oil spill monitoring and satellite and sensor composition for fire control

3.   business planning and outsourcing to new countries, and in particular supporting the choice of locations and business plan. .

Details of the applications and the specific chosen scenarios are given in [5].

We have chosen to focus on e-business applications for four main reasons.

Firstly, and more importantly, although there are similarities between (conventional) scientific grids and business-oriented grids, there is a substantial distance in existing grid technology that needs to be filled for the competitive realisation applications making use of the potential offered by the grid while at the same time being appealing to users and providers, and this is particularly evident for e-business applications. Indeed, e-business requires sharing computing power, databases, instruments, and other on-line tools securely across corporate, institutional, and geographic boundaries without sacrificing local autonomy, as allowed by a grid infrastructure, where the following features, important to e-business, exist:

- continuous access of the data and services to process data under request is provided;
- real-time data is allowed, particularly for commodity item prices, exchange rates, availability of quantities and delivery dates
- accurate data delivery is ensured, importantly for the final applicability of the service in the decision process of the buyer-user;
- new data sources may be included and new process to be added;
- data and computation are intensive;
- the solution is robust to failure.

Secondly, we believe that grid technology in general and ArguGRID in particular will be highly beneficial to the e-business community and, in the long term, contribute to substantially improve the competitiveness of European business. Indeed, virtual organizations of argumentative agents will be able to emulate, to a large extent, the way Buyers and Suppliers interact in the real world, but in a more efficient and effective manner. More specifically, from the Buyer's side, if we take into account:

- the difficulty he has today in gathering quickly and timely, relevant and accurate information to satisfy a specific -but not detailed yet-, procurement need,
- the necessity to keep refining and possibly redefining his search, as he gains more information and knowledge
- the need to narrow down the number of suppliers (or groups of suppliers in complex cases) and to develop the "short list"
- and finally, the need to negotiate the best deal, between the selected and theoretically equivalent (quality-wise) suppliers,

then we conclude that an agents/VO/ argumentation model will advance the way towards a less deficient market .

However, a semantic layer is needed in order to realise business applications, allowing for the high-level specification for information and services, given well defined and explicitly represented meaning, better enabling computers and people to work in cooperation. One way of achieving this goal is to apply existing Semantic Web technologies (metadata standards, reasoning engines...) to Grid computing. Another way is to deploy agents, suitable for e-business application[1] as they can aid decision making by users and can configure naturally in agent societies to provide business on demand such as e-business. They can also naturally hold representations of and reason with users' requirements and service descriptions and requirements and thus provide customized solutions, as required by e-business. Finally, they can engage in extensive negotiation, almost always necessary before cooperation as required by e-business can take place, on behalf of the users and service providers, and form agents societies (VOs). Thus, in order to accommodate e-business applications, we will benefit the grid by providing a semantic layer to it, comprising of semantic web technologies, argumentative agents, and virtual organisations. In other words, these applications will be a vehicle for enriching the grid.

Thirdly, because of the reasons for deploying agents and semantic web technologies given above, we believe that e-business applications are highly suitable to illustrate the functioning and benefits of the ARGUGRID approach to grid computing. In an e-business application, there are several autonomous players who will act only to safeguard their own interests but who are willing and looking for cooperation to increase their wealth. Cooperation will almost always be preceded by a negotiation phase in which the participants have to agree on what to do and how to do it and how to share the benefits. Further the participants in such negotiation do not necessarily know each other in advance. This represents an appropriate case for the application of virtual organisations composed of autonomous agents, where several issues need to be better understood: negotiation, dispute resolution, trust, and stability. Moreover, for e-business, web services are already acknowledged as important, and we plan to make use of these too.

Finally, we believe that the realisation of models and infrastructure for these applications will have far reaching benefits, and plan to explore these benefits for (conventional) grid-enabled earth observation applications.

Acknowledgments. This work was partially funded by the Sixth Framework IST programme of the EC, under the 035200 ARGUGRID project. Many thanks to all participants in the ArguGRID consortium, and in particular Thanassis Stournaras and Dimitris Dimitrelos, for helpful comments on an earlier version of this paper.

References

1. Curcin, V., Ghanem, M., Guo, Y., Stathis, K., Toni, F.: Building next generation Service-Oriented Architectures using argumentation agents. In: Proc. 3rd International Conference on Grid Services Engineering and Management (GSEM 2006) (September 2006)
2. Dung, P.M., Mancarella, P., Toni, F.: Computing ideal sceptical argumentation. Artificial Intelligence, Special Issue on Argumentation in Artificial Intelligence (to appear in September/October 2007)

---

[1] Computational Intelligence, special issue on Business Agents and the Semantic Web (BASeWEB), 2004.

3. Gaertner, D., Toni, F.: CaSAPI: a system for credulous and sceptical argumentation. In: Proc. LPNMR-Workshop on Argumentation and Non-Monotonic Reasoning (ArgNMR07) (May 2007)
4. Morge, M., Mancarella, P.: The hedgehog and the fox. An argumentation-based decision support system. In: Proc. Fourth International Workshop on Argumentation in Multi-Agent Systems (ArgMAS 2007) (May 2007)
5. Stournaras, T., Dimitrelos, D., Tabasco, A., Barba, J.A., Pedrazzani, D., Yagüe, M.J., An, T.C., Dung, P.M., Hung, N.D., Khoi, V.D., Thang, P.M.: e-Business application scenarios. ArguGRID deliverable D.1.2 (May 2007)
6. McGinnis, J., Bromuri, S., Urovi, V., Stathis, K.: Automated Workflows Using Dialectical Argumentation. In: German e-Science Conference, Germany (to appear, 2007)
7. Stathis, K., Lekeas, G.K., Kloukinas, C.: Competence checking for the global e-service society using games. In: Engineering Societies in the Agents World (ESAW06), Springer, Heidelberg (to appear, 2007)
8. Modgil, S., McGinnis, J.: Towards Characterising Argumentation Based Dialogue in the Argument Interchange Format, ArgMAS 2007, Hawai US (to appear, 2007)
9. Stathis, K., Kafetzoglou, S., Pappavasiliou, S., Bromuri, S.: Sensor Network Grids: Agent Environment combined with QoS in Wireless Sensor Networks. In: Proceedings of the 3rd International Conference on Autonomic and Autonomous Systems (ICAS'07), Athens, GR (to appear, 2007)
10. Miller, T., McBurney, P., McGinnis, J., Stathis, K.: First-Class Protocols for Agent-Based Coordination of Scientific Instruments. In: Proceedings of the 5th International Workshop on Agent-based Computing for Enterprise Collaboration: Agent-Oriented Workflows and Services (ACEC 07), Paris, France (to appear, 2007)

AssessGrid, Economic Issues Underlying Risk Awareness in Grids*

Kerstin Voss[1], Karim Djemame[2], Iain Gourlay[2], and James Padgett[2]

[1] Paderborn Center for Parallel Computing, University of Paderborn, Germany
kerstinv@uni-paderborn.de
[2] School of Computing, University of Leeds, United Kingdom
{karim,iain,jamesp}@comp.leeds.ac.uk

Abstract. In order to improve the attractiveness and drive the commercial uptake of Grid technologies, the establishment of Service Level Agreements (SLAs) is required. The AssessGrid project contributes to this aim by introducing risk-aware Grid architectural components. Grid service users, brokers and providers benefit from risk assessment functionalities in all phases of service provisioning and utilisation. This paper focuses on the economic issues which result from this new risk-aware approach to Grid computing. Multiple open economic research questions are discussed from the perspective of users, brokers and providers, which point out the potential impact of AssessGrid in this area.

Keywords: Risk Assessment, Risk Management, SLA, AssessGrid.

1 Introduction

The AssessGrid project is motivated by the need to bridge the gap between Service Level Agreements (SLAs) as a concept and as an accepted tool in Grid utilisation and service provisioning. If widespread acceptance is to become a reality, a number of issues need to be addressed that are important to the two main Grid actors: the Grid service providers and the Grid service users. Service providers may be unwilling to agree an SLA since they are aware of the possibility of resource failures. This could lead to an SLA violation and consequently a need to pay a penalty fee. Meanwhile, end-users do not completely trust SLAs since they are also aware of possible resource failures and know that a *one hundred percent* guarantee cannot be given by any provider. To establish the usage of SLAs for Grids, which is essential for commercial Grid utilisation, the aim of the AssessGrid project [1] is to integrate a risk-aware SLA model into current Grid technology. The provision of risk assessment methods brings added benefits for end-users as well as a new potential market for Grid services and resource brokers. The risk information is integrated in SLAs as an additional negotiable parameter in order to notify end-users about the probability that the SLA may

---

* This work has been partially supported by the EU within the 6th Framework Programme under contract IST-031772 "Advanced Risk Assessment and Management for Trustable Grids" (AssessGrid).

D.J. Veit and J. Altmann (Eds.): GECON 2007, LNCS 4685, pp. 170–175, 2007.

be violated. The probability of failure (PoF) published in the SLA will enable end-users to compare different SLA offers. In particular, the PoF will influence the price and the penalty offered by the Grid service provider, e.g. high costs and penalties for job executions which have a high predicted success rate and vice versa. Through this extension end-users obtain a new perspective when selecting an SLA offer since they can individually evaluate the balance between the consequences of an SLA violation and the price they are willing to pay for Grid service usage. The use of PoFs in SLAs opens new possibilities for Grid brokers in the scope of workflows and the evaluation of the reliability of PoF values published by providers.

In order to enable Grid service providers to offer a PoF for an SLA violation, risk assessment methods have to be integrated into the Grid fabric. The usage of risk information is, however, not limited to publishing it in SLA offers. Rather the AssessGrid developments integrate risk management methods for brokers and providers which use the assessed risk as a decisive component in their scheduling and resource management activities.

This paper presents an economic view of the AssessGrid project in order to point out the potential impact of the project. It is structured as follows. In section 2 an overview of the project is given. In section 3 the identified Grid economy research issues are described. General conclusions and future work are presented in section 4.

2   AssessGrid: Risk Management in the Grid

AssessGrid aims to introduce risk assessment and management [2] in a Grid environment. In professional risk management, risk is not only the likelihood of occurrence. Since in the scope of risk management different events have to be compared, it is not sufficient to only consider their likelihood. To develop an accurate risk management process, the consequences of an event also have to be taken into account when comparing different threats. Accordingly, a more complete definition describes risk in terms of the product of the likelihood of an event and the impact of its occurrence. From an economic perspective, it is desirable to express the impact in monetary terms. However, a provider or broker cannot, in general, evaluate the financial impact of an SLA violation on an end-user since the data required to make such an evaluation is not available to them. Consequently, the provider offers a probability of failure (PoF) instead of a risk value. Note that the risk for an SLA violation from the provider's perspective can be determined exactly by the PoF and the agreed penalty.

We have identified three main actors in the scope of Grid service provisioning and utilisation: end-user, broker, and provider. The end-user is the Grid service consumer, who specifies quality of service (QoS) requirements through an SLA. In the computational Grid, end-users run various applications and define SLAs on a per-application basis since each Grid job has different requirements and time constraints. The Grid fabric is comprised of resources that are owned by various Grid resource providers. Providers charge for the utilisation of their resources, agreeing SLAs that specify the cost, the provider's obligations, and penalty fees to be paid in the event that those obligations are not met. Grid resources can then either be accessed directly or through the use of a broker. A broker is defined as a business role that acts as a third party between organisations that consume Grid resources and Grid providers that offer these services. It can offer additional services to the end-user due to benefiting from the data acquired through its many negotiations with various providers.

Details on these actors as well as the architecture for the risk-aware Grid components developed in the project can be found in [3].

3   Grid Economy Research Issues

The objective of AssessGrid is to contribute to the establishment of commercial adoption of Grid technology. Therefore it is essential to introduce economy-awareness into each role - end-user, broker, and provider. This sections presents aspects of the architecture where an economy model may be exploited. The current scope of AssessGrid is the provision of composite (e.g. SLAs, workflows etc) and computational (e.g. physical Grid infrastructure) service markets.

3.1   End-User

The end-user is provided with a number of abstract applications which make use of Grid services deployed within the Grid fabric layer. SLA requests and offers are exchanged between end-user and broker or provider, in order to agree an SLA which grants permission to invoke a Grid service in the fabric layer. Within each layer, the organisation performing the role of each actor must define a policy statement governing the acceptable bounds of negotiation. This restricts end-users and contractors to request or offer SLAs which fall outside of the organisation's acceptable limits. For example, in addition to specifying budget constraints, there may be a restriction on a provider's penalty conditions to limit the financial loss incurred because of an SLA violation. Taking these policy limits into consideration, an end-user can negotiate an SLA to run a Grid service by defining requirements as well as the requested QoS in an SLA request. During the definition process the end-user evaluates the importance of the job in terms of its urgency and the consequences of delayed results or failure. A further validation of the policy limits must be made against the SLA offers received from the broker or providers.

Where several SLA offers have been negotiated on behalf of the end-user, the broker can return a ranked list - according to price, penalty, and PoF. The challenge for the end-user is to find an SLA offer which offers the best service in terms of price, penalty, and PoF. Here we can apply a mathematical model to help the end-user make the *best* offer selection based on quality criteria. The end-user defines a ranking of the quality criteria (e.g. PoF is more important than price) in order to measure each of the offers according to its closeness to the criteria. A possible approach is the application of an Analytic Hierarchy Process which is based on criteria weights specified by the end-user [4].

From the perspective of the end-user, there are economic issues if the SLA is violated. In this case the end-user needs to evaluate whether the penalty received is sufficient to offset the consequences of the SLA violation. The evaluation results will be useful information for the negotiation of similar SLAs in the future.

3.2   Broker

Within the AssessGrid architecture the broker role facilitates SLA negotiation between entities fulfilling the end-user and resource provider roles. After the negotiation has returned an SLA offer, the broker is responsible for performing reliability checks on the PoFs contained in the SLA offers. Without this check, the end-user has no independent view on the provider's assessment, which cannot be assumed to be impartial. SLA offers that are deemed to be unreliable are subjected to an additional risk assessment by the broker using historical data related to the provider making the offer. Where multiple SLA offers are returned by the SLA negotiation process, the broker can rank these according to a price, penalty, PoF matrix depending on the priorities of the end-user.

The economic benefit of using a broker within the SLA negotiation process effects all three Grid actors and provides the opportunity for an economy model where SLAs for software services are bought and sold based on differentiated classes of service. In the provision of SLA negotiation, the broker offers two classes of service: mediator and runtime-responsible. In the case of mediator service, the broker provides a marketplace for providers to advertise their SLA templates to a wider number of end-users; for end-users it allows selecting a provider and their services from a larger set.

Use of the broker as a runtime-responsible service offers the greatest scope for Grid economy research. In this scenario the broker has the ability to buy SLA offers from providers and resell them to end-users transparently using its own SLA offer. In this way a broker becomes a virtual provider and can offer price, penalty, and PoF combinations unavailable from a single provider. In addition, the broker can orchestrate workflows, which combine multiple single SLA offers and combine these into SLA offers for an entire workflow. The broker can make trade-offs against price, penalty, and PoF between providers in order to maximise the economic benefit for itself. Where a task is executed redundantly, to reduce the PoF, or where it forms part of a workflow orchestration, the broker has additional responsibilities during runtime. Should an SLA governing one of these tasks be violated, the broker must determine whether it is more economical to pay its own penalty fee or to negotiate a new SLA offer with a different provider at a price which minimises its losses.

During post-runtime, the broker is responsible for updating the historical data it holds on each provider registered therewith. When offering it's runtime-responsible service, the broker can easily access the final status of SLA offers as they are agreed between itself and the provider. When acting as a mediator, the broker must persuade the end-user to pass on the same information about the SLA final status. A rebate or bonus payment may be built into the economy model to encourage end-users to give SLA offer feedback to the broker. As well as the financial benefit, end-users will benefit through up-to-date historical data and greater confidence in the reliability checks.

3.3   Resource Provider

A provider offers access to resources and services through formal SLA offers specifying the requirements as well as PoF, price, and penalty. Providers need well-balanced infrastructures, so that they can maximise the offerable QoS and minimise the number of SLA violations. Such an approach increases the economic benefit and motivation of end-users to outsource their IT tasks.

A number of economic issues have been identified which affect the provider. These issues can be categorised as belonging to the pre-runtime (i.e. during SLA negotiation), run-time and post-runtime phases.

In the pre-runtime phase a risk aware negotiation requires that a provider place an advance reservation for the SLA and calculates the PoF [5]. Based on this, a provider determines the price and penalty fee which will be offered to an end-user. To ensure unsuitable SLA offers are not made, end-users define minimum and maximum limits for price, penalty, and PoF within the SLA request. A provider's decision whether to agree or reject an SLA depends on the fees and the requested PoF in comparison with the current status of its infrastructure.

The publication of the SLA PoF opens further research fields. A provider must not offer the PoF it had assessed during the reservation process since no mechanism can be developed which can coerce it into telling the truth. However, the broker's confidence service is designed to ensure that providers do not lie about published PoFs. Therefore it is the ability of the provider to fulfill SLA offers which marks it out as reliable, rather than its ability to offer SLAs with a low PoF.

For contractors (end-users or brokers), an important provider selection criterion is the price. The SLA template contains pricing information for actions such as data transfer, CPU usage, and storage. Within the AssessGrid model these prices are variable since the price depends on the PoF value specified within the SLA.

The market mechanism will influence the pricing since each provider has only a limited resource set with variable utilisation. Consequently, prices for resource usage will not be fixed but will depend on the economics of supply and demand. Reservations, which were made well in advance, will usually result in a reduced price since there will be access to a greater number of free reservation slots. Equally, immediate resource usage may also result in reduced prices, as providers try to increase their utilisation if demand is low. However, end-users risk resources unavailability if they wait too long before reserving resources. These pricing dynamics are valid only in the scope of the resource costs and do not consider the PoF.

After an SLA has been agreed by the provider and the end-user, the provider has to ensure during runtime that the SLA will not be violated. Accordingly, the provider's risk management activities are controlled by estimating the penalty payments in the case of an SLA violation. Using the AssessGrid technology enables to initiate precautionary fault tolerance mechanisms in order to prevent SLA violations. The penalty fees, in addition to the PoF (i.e. risk), are the decisive factors in determining which fault tolerance mechanisms are initiated.

In the post-runtime phase the provider has to evaluate the final SLA status to determine whether a penalty fee has to be paid. Even in the case the SLA had been fulfilled the costs for the fulfillment have to be checked since the initiation of a fault tolerance mechanism also consumes resources and therewith results in additional costs. The results of the evaluation process will point out on the one hand whether adjustments in the offer making policies are necessary in order to increase the provider's profit. On the other hand, statistics can be generated which show whether initiated fault tolerance mechanisms had been able to prevent an SLA violation.

4   Conclusion

The integration of risk-awareness into the Grid provides a number of benefits within an economy framework but also gives rise to numerous research problems. This paper presents an overview of the AssessGrid developments and their impact for continuing research in economic models. An unexplored problem is the handling of workflows for a broker. Since the broker is responsible for the SLA fulfillment, it has to react on failures (negotiate with providers for a repeated job execution) in order to prevent paying penalties. Other essential economic issues of AssessGrid are the pricing mechanisms for brokers and providers which must take account of the probability of failure in a risk-aware Grid approach. Further research is required to address these problems in detail.

References

1. AssessGrid : (Advanced risk assessment and management for trustable grids), http://www.AssessGrid.eu
2. Koller, G.: Risk Assessment and Decision Making in Business and Industry. CRC Press, Boca Raton (1999)
3. Birkenheuer, G., Hovestadt, M., Voss, K., Kao, O., Djemame, K., Gourlay, I., Padgett, J.: Introducing Risk Management into the Grid. In: Proceedings of the 2nd IEEE International Conference on e-Science and Grid Computing, Amsterdam, The Netherlands (2006)
4. Saaty, T.: How to make a decision: The Analytic Hierarchy Process. European Journal of Operational Research 48, 9–26 (1990)
5. Hovestadt, M., Kao, O., Voß, K.: The First Step of Introducing Risk Management for Prepossessing SLAs. In: IEEE International Conference on Services Computing (SCC), Chicago (2006)

CATNETS – Open Market Approaches for Self-organizing Grid Resource Allocation

Torsten Eymann, Werner Streitberger, and Sebastian Hudert

Bayreuth University, 95447 Bayreuth, Germany
torsten.eymann@uni-bayreuth.de
http://www.wi.uni-bayreuth.de/

Abstract. Grid computing has recently become an important paradigm for managing computationally demanding applications, composed of a collection of services. The dynamic discovery of services, and the selection of a particular service instance providing the best value out of the discovered alternatives, poses a complex multi-attribute n:m allocation decision problem. Decentralized approaches to this service allocation problem represent a flexible alternative to central resource brokers, thus promising improvements in the efficiency of the resulting negotiations and service allocations. This paper analyses the impact of the service density on the profit and market price estimation using a decentralized service allocation mechanism in a grid market scenario.

Keywords: Self-Organisation, Economic Resource Allocation, Grid Service Allocation.

1 Introduction

Grid computing represents a concept for coordinated sharing of globally distributed resources spanning several physical organizations [1]. Currently the idea of Service-Oriented Architectures (SOAs) underlie several of the current Grid initiatives and reflect the common approach to realize Grid computing infrastructures, where participants offer and request application services. SOA defines standard interfaces and protocols that enables developers to encapsulate resources of different complexity and value as services that clients access without knowledge of their internal workings [2]. Grid systems have therefore increasingly been structured as networks of inter-operating services that communicate with one another via standard interfaces. Such infrastructures of services provided to an a priori unknown set of consumers can be efficiently organized as markets, analogously to traditional service markets in real world economies. Grid computing can thus become an object of Economics research, and thus provide insights not only for computer scientists, but also for economists.

The design and construction of resource allocation schemes is a particular research topic that can be tried (and evaluated) in globally distributed, large-scale Grid environments. Apart from computable general equilibrium approaches (NP-complete and thus not feasible) and all kinds of auctions (a Grid eBay?),

D.J. Veit and J. Altmann (Eds.): GECON 2007, LNCS 4685, pp. 176–181, 2007.

it becomes also possible to investigate in self-organization approaches. Self-organization can be found everywhere in our world, e.g. biological evolution, social group behaviour, market dynamics phenomena and other complex adaptive systems.

This article describes an investigation in implementing a self-organizing Grid Market based on the "'Catallaxy"' concept of F. A. von Hayek [3]. Catallaxy describes a "'free market"' economic self-organization approach for electronic services brokerage, which can be implemented for realizing service markets within service-oriented grid computing infrastructures. In such infrastructures, participants offer and request actual application services and computing resources for providing such services, of different complexity and value - creating interdependent markets:

- a service market - which involves trading of application services, and
- a resource market - which involves trading of computational and data resources, such as processors, memory, etc.

The distinction between resource and service allows different instances of the same service to be hosted on different resources. It also enables the price of a given service to base on the particular resource capabilities that are being made available by the hosting environment. In such interrelated markets, allocating resources and services on one market inevitably influence the outcome on the other market. This concept of two interrelated markets takes the current Grid concept one step further.

This paper investigates the general outcome of decentral resource negotiations in Grid systems. For this purpose a particular Grid environment is implemented and used for the actual simulation runs. Using Grid simulation software, different economic settings are investigated. The simulation results are evaluated using a defined set of metrics. The paper concludes with discussing the resulting metrics.

2   Related Work

The use of market mechanisms for allocating computer resources is not a completely new phenomenon. Regev and Nisan propose within the scope of the POPCORN project the application of a Vickrey auction for the allocation of computational resources in distributed systems [5].

Buyya motivated the transfer of market-based concepts from distributed systems to Grids [6]. However, he proposed classical one-sided auction types, which cannot account for combinatorial bids. Wolski et al. compare classical auctions with a bargaining market [7]. As a result, they come to the conclusion that the bargaining market is superior to an auction based market. Eymann et al. introduce a decentralized bargaining system for resource allocation in Grids, which incorporates the underlying topology of the Grid market [8].

Subramoniam et al. account for combinatorial bids by providing a *tâtonnement* process for allocation and pricing [9]. Wellman et al. model single-sided auction protocols for the allocation and scheduling of resources under consideration of different time constraints [10]. Conen goes one step further by designing a combinatorial bidding procedure for job scheduling including different running, starting, and ending times of jobs on a processing machine [11]. However, these approaches are single-sided and favor monopolistic sellers or monopsonistic buyers in a way that they allocate greater portions of the surplus. Installing competition on both sides is deemed superior, as no particular market side is systematically put at advantage.

3   Simulation Model

This section describes the Grid simulation model used to simulate the Catallactic free-market allocation approach. The CATNETS Grid simulator – an extension of the OptorSim Grid Simulator [15]– is used for simulation. The Grid network (GN) is defined by a connected non-oriented graph

$$GN = \langle S, L \rangle$$

with $S = n$ network sites and L a set of links which connect the sites with a bandwidth. The BRITE network generator is used to create the links between the sites [14]. Each site is characterized by a triple $\langle CSA_i, BSA_i, RA_i \rangle$ where CSA_i is a set of *Complex Service Agents (CSAs)*, BSA_i is a set of *Basic Service Agents (BSAs)*, and RA_i is a set of *Resource Agents (RAs)*. In every site there can be zero or more complex/basic service agents and zero or more resource agents. A node with no associated agents is a *router*.

Complex Service Agents. CSAs are entry points to the Grid system and are able to execute *Complex Services (CSs)* for Grid clients. A CS is defined as a set of *Basic Services (BSs)*. CSAs are not specialized: they accept any type of complex service request and take care of the execution of the component basic services. For simplicity reasons, a complex service requests always one basic service in the evaluated scenario. Several CSAs are available in the network, which enable parallel allocation and execution of BSAs.

Basic Service Agents. BSAs provide CSAs with the BSs they need to furnish their complex services to Grid clients. A predefined number of BSAs is available for selection of the CSAs.

Resources Agents. Resources have a *name* which is a unique identifier whose intended semantics is shared among all agents. Every resource is also characterized by a *quantity* whose value is a positive integer. RAs are "proxies" for aggregations of resources. Their task is to provide BSAs with resources needed for the execution of basic services. For simplicity reasons, a RA provides only of one unit of their resource and the BSA requests one unit from RA.

4   Simulation Scenario

The simulation scenario analyses the impact of the agent's density on the outcome of the market. The total number of agents changes between the simulation scenarios from 30 and 60 to 300, while keeping the number of Grid sites fixed. In detail, the number of agents is equally split in CSAs, BSAs and RAs. For example, a total number of 60 agents means a set of 20 CSAs, 20 BSAs and 20 RAs. The agents are distributed over the 30 Grid sites using a uniform distribution. The bandwidth between the sites is set to 1 Gbit/s, which guarantees similar communication conditions.

All agents' price intervals are initialized in an interval between 80 and 180, with an interval length of 30. The lower bound for BSAs is drawn from a uniform distribution. Its upper limit is obtained as a simple addition of 30 to the lower limit value. The initial price interval for the CSAs is computed drawing a number from the random price interval and a subtraction of 30.

A CSA requests a basic service by broadcasting a call-for-proposal message to BSAs, which is received by all BSAs reachable within 2 hops in the given network topology.After a discovery timeout of 500 ms, the requesting CSA selects one BSA for negotiation. A best price selection policy picks the best offered proposal and starts the negotiation by iterative and bilateral message exchange, until one party accepts or rejects.

5   Evaluation

The simulation scenario is evaluated with two metrics on the population level. Figure 1 shows the profit and the estimated market price of the complex service and basic service agents during one simulation run in different settings. The profit is computed as the difference between negotiation price and the estimated market price. The goal of the agents is to optimize their profit. The estimated market price is computed as a weighted average of historical agreement prices.

The analysis concentrates on the evaluation of the service market; similar results are found at the resource market. In each simulation run, 1000 complex services issues requests, the delay between the requests was 1000 ms and the execution time of one basic service was set to a constant 1000 ms. Each diagram shows the BSAs as service providers and CSAs as service consumers for a population of 30, 60 and 300 agents.

In the smallest scenario, the profit of the agents converges fast to values near zero. Both, complex services and basic services trade very often and estimate the market price very well. They use their good market price estimation to optimize their trading strategy. None of the trading partners is able to make a large profit. Only at the beginning of the simulation the profit/loss peaks indicate wrong market price estimations due to insufficient knowledge on market prices. This effect increases with 60 agents and leads to high deviations in the 300 agent scenario, as the necessary amount of information needed for feedback learning decreases *per capita*. Only sellers are able to make a profit - they capitalize on the (far too low) market price estimations of the complex services and realize high profits. At the end of the simulation run, the buyers slightly increase their profits due to better market price estimations. The number of 1000 requests is not enough for all agents to learn the market price and optimize their behavior within a population of 300 agents.

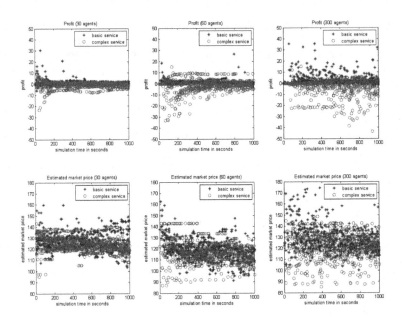

Fig. 1. Profit and estimated market price for 1000 requests and 30, 60 and 300 agents

6 Conclusion and Outlook

The paper presented an evaluation of a Catallactic free-market strategy, where no central resource broker exists. Different scenario settings are evaluated dependent on the service/agent density. In scenarios with a high number of agents, the Catallactic strategy is slow in converging to a stable market price estimation and profit of the market participants, because the individual agents interact not enough to arrive at a stable market price estimation. Further experiments increasing the number of requests have to be performed to analyze its impact on the agent's profit.

Acknowledgements

This work has partially been funded by the EU in the IST programme "Future and Emerging Technologies" under grant FP6-003769 "CATNETS".

References

1. Foster, I., Kesselmann, C.: The Grid: Blueprint for a New Computing Infrastructure. Morgan Kaufmann Publishers Inc. San Francisco, CA (1999)
2. Foster, I.: Service-Oriented Science. Science Journal 308, 814–817 (2005)
3. von Hayek, F.A.: The Use of Knowledge in Society. American Economic Review XXXV 4, 519–530 (1945)

4. Eymann, T., Sackmann, S., Mueller, G., Pippow, I.: Hayek's Catallaxy: A Forward-looking Concept for Information Systems. In: Proc. of American Conference on Information Systems (AMCIS'03) (2003)

5. Regev, O., Nisan, N.: The POPCORN Market: Online Markets for Computational Resources. In: Decision Support Systems, vol. 28, pp. 177–189. Elsevier Science Publishers, Amsterdam (2000)

6. Buyya, R., Stockinger, H., Ghiddy, J., Abramson, D.: Economic Models for Management of Resources in Peer-to-Peer and Grid Computing. In: Proceedings of the International Conference on Commercial Applications for High Performance (2001)

7. Wolski, R., Brevik, J., Plank, J.S., Bryan, T.: Grid Resource Allocation and Control using Computational Economies. In: Berman, F., Fox, G., Hey, A. (eds.) Grid Computing: Making The Global Infrastructure a Reality, pp. 747–772. John Wiley & Sons Publishers, West Sussex (2003)

8. Eymann, T., Reinicke, M., Ardaiz, O., Artigas, P., de Cerio, L.D., Freitag, F., Messeguer, R., Navarro, L., Royo, D.: Decentralized vs. Centralized Economic Coordination of Resource Allocation in Grids. In: Proceedings of the 1st European Across Grids Conference (2003)

9. Subramoniam, K., Maheswaran, M., Toulouse, M.: Towards a Micro-Economic Model for Resource Allocation in Grid Computing Systems. In: Proceedings of the 2002 IEEE Canadian Conference on Electrical & Computer Engineering (2002)

10. Wellman, M., Walsh, W., Wurman, P., MacKie-Mason, J.: Auction Protocols for Decentralized Scheduling. Games and Economic Behavior 35, 271–303 (2001)

11. Conen, W.: Economically Coordinated Job Shop Scheduling and Decision Point Bidding - An Example for Economic Coordination in Manufacturing and Logistics. In: Proceedings of the 16th Workshop on Planen, Scheduling und Konfigurieren, Entwerfen (2002)

12. Parkes, D., Kalagnanam, J., Eso, M.: Achieving Budget-Balance with Vickrey-Based Payment Schemes in Exchanges. In: Proceedings of the Seventeenth International Joint Conference on Artificial Intelligence, pp. 1161–1168 (2001)

13. Biswas, S., Narahari, Y.: An Iterative Auction Mechanism for Combinatorial Logistics Exchanges. In: Proceedings of the 9th International Symposium on Logistics (2004)

14. BRITE: Boston university Representative Internet Topology gEnerator, http://www.cs.bu.edu/brite/

15. OptorSim: http://sourceforge.net/projects/optorsim

The edutain@grid Project*

Thomas Fahringer[1], Christoph Anthes[2], Alexis Arragon[3], Arton Lipaj[4],
Jens Müller-Iden[5], Christopher Rawlings[6], Radu Prodan[1], and Mike Surridge[7]

[1] Institute for Computer Science, University of Innsbruck, Austria
{tf,radu}@dps.uibk.ac.at
[2] Institute of Graphics and Parallel Processing, University of Linz, Austria
canthes@gup.jku.at
[3] Darkworks S.A., France
A.Arragon@darkworks.com
[4] Amis d.o.o., Slovenia
arton.lipaj@amis.net
[5] Institute of Computer Science, University of Münster, Germany
jmueller@math.uni-muenster.de
[6] BMT Cordah Ltd., U.K.
chris.rawlings@bmtcordah.com
[7] IT Innovation Centre, University of Southampton, U.K.
ms@it-innovation.soton.ac.uk

Abstract. edutain@grid is an exciting and ground breaking new project
making use of Grid technology. The project will identify and define a new
class of applications that are highly significant for Grid computing but
have not been studied in the past, which we characterise as Real-Time
Online Interactive Applications (ROIA). The distinctive features that
make ROIA unique include large user concurrency to a single application instance, ad-hoc connections, competition-oriented Virtual Organisations, real-time interactive response, dynamically changing control and
data application flows whilst maintaining high Quality of Service (QoS),
user friendly security, and novel Business-to-Consumer market models.
In order to meet these challenges, the project team will develop a new
middleware layer that will allow ROIA to exploit Grid computing and
validate the system using two pilot applications from online gaming and
e-learning domains.

Keywords: Business models, E-learning, Grid computing, Online games,
Quality of Service, Real-time Online Interactive Applications, Service
Level Agreement, Scalability.

1 Introduction

For some years, Grid computing [1] has been successful in certain academic research disciplines, allowing researchers to share their computational resources or

---

* This research is supported by the European Union through the IST-034601 edu-
tain@grid project.

D.J. Veit and J. Altmann (Eds.): GECON 2007, LNCS 4685, pp. 182–187, 2007.

data to achieve an agreed research goal that none could pursue on their own. In industry, Grid computing has also been partially successful to improve time to market through rapid deployment of resources for new projects in areas such as pharmaceutical industries and risk analysis of financial services. However, Grid technology has yet to make an economic or societal impact similar to that achieved in the last 15 years by Web technologies. There are many reasons for this, including the lack of convergence of underlying specifications, the economic cost of porting applications, the inaccessibility of Grids from a usability perspective, especially with respect to Grid and Virtual Organisation security models, the high economic cost and static nature of Grid deployments and operation, the limited support for guaranteed performance, scalability, failover and recovery, or the lack of support for business models that provide an attractive balance of risks and rewards for both providers and consumers of services.

Possibly the most important (but rarely mentioned) reason for the slow pace of progress compared with Web technologies is the lack of any obvious "killer" applications for the Grid. In this project, we identify and define a new class of applications that are highly significant for Grid computing but have not been studied in the past, which we characterise as *Real-Time Online Interactive Applications (ROIA)*.

We classify ROIA as a new class of Grid applications with the following distinctive features that makes them unique in comparison to traditional parameter study or scientific workflows, highly studied by previous Grid research [2]:

- The applications often support a very large number of users connecting to a single application instance;
- The users sharing an application interact as a community, but they have different goals and may compete (or even try to cheat) as well as cooperate with each other;
- Users connect to applications in an ad-hoc manner, at times of their choosing, and often anonymously or with different pseudonyms;
- The applications mediate and respond to real-time user interactions, and typically involve a very high level of user interactivity;
- The applications are highly distributed and highly dynamic, able to change control and data flows to cope with changing loads and levels of user interaction;
- The applications must deliver and maintain certain QoS parameters related to the user interactivity even in the presence of faults.

The distinctive features that make ROIA unique include large user concurrency to a single application instance, ad-hoc connections, competition-oriented Virtual Organisations, real-time interactive response, dynamically changing control and data application flows whilst maintaining high QoS, user friendly security, and novel Business-to-Consumer market models. In order to overcome these challenges, the project team will develop a new middleware layer that will allow ROIA to exploit Grid computing and validate the system using two pilot applications from online gaming and e-learning domains.

2   Objectives

Grid technology still does not provide good support for all key features required by ROIA. The performance overheads of current Grid protocols and logically centralised resource management act against real-time interactivity, for example. Traditional Grids also do not provide good support for sharing application instances among communities of users with different (possibly conflicting) goals from each other or from resource providers. The edutain@grid project therefore seeks to overcome these barriers and implement Grid-based ROIA by having the following key scientific and technical objectives:

- To define ROIA as a new class of socially important applications and provide complete Grid support for their key features through well-chosen sample applications;
- To provide a QoS-enabled middleware for negotiation of Service Level Agreements (SLA) and ROIA provisioning that copes with dynamic Grid and highly populated user environments;
- To develop mechanisms that provide the necessary real-time performance, scalability, manageability, and QoS;
- To devise business models that make the provision of large-scale ROIA economically viable;
- To make the Grid accessible to large numbers of users of such applications by overcoming usability barriers including those associated with Grid security;
- To make the resulting technology cost-effective for application developers;
- To disseminate and promote exploitation and take-up of the technological results.

3   Pilot Applications

We consider many different classes of applications to be ROIA, like online games, online e-learning environments, training simulations, or synchronous collaborative work environments (engineering and science). The sample applications chosen to validate the scientific and technological developments are in multi-player online gaming and e-learning. By targeting these socially important sectors (education and entertainment), edutain@grid seeks to accelerate the emergence of killer Grid applications and promote accelerated take-up of the technology by European business and society.

ROIA are characterised by the tight immersive coupling of users to the application, the high rate of interactions between users and the frequent state computation and communication participating computers over the Internet. In some very responsive action computer games, the distributed processes exchange new application information at a very high rate of up to 35 updates per second. Users immediately notice a delay in this distributed computation and communication as a "lag" in the interactive flow and their immersion is abruptly disturbed. Because of this tight coupling of distributed processes, current ROIA run in a static way and do not allow dynamical adding, removing or migrating of the used resources.

Similarly the benefits to the e-learning community are expected to be significant where large numbers of geographically disparate students can interact with instructors making use of large operational data sets. In particular this will be relevant to online simulations in scientific modelling applications used in the energy, defence, transport and legal market sectors. Furthermore, edutain@grid is expected to attract new developers and development ideas that were not previously possible or simply cost prohibitive. One metric of the project being the generic support for ROIA, edutain@grid system will not be limited to the two domains of pilot applications. However, both online games and e-learning domain represent large classes of applications: online games as covered by edutain@grid refer to first person shooter games, action and adventure games and the game pilot application will only explore one of those. We will examine multiple scenarios by varying number of users and sessions, thus exploring in depth the nature of these applications and will provide tools and product open enough to support a wide variety of application domains, thus verifying the generality of edutain@grid.

4   Products

The outcome of the project will comprise the following elements:

- A *business infrastructure* supporting business models that make the provision of large-scale ROIA economically viable;
- A *management infrastructure* which handles the dynamic execution of ROIAs on the Grid with support for advanced configuration management;
- A *runtime framework* [3] which enables scalability and advanced Grid functionality within real-time applications.

These three elements address specific needs of the different user classes and their respective key challenges.

4.1   Client Products

The project will produce a so called client manager, providing a light-weight secure client application for distributing and installation a ROIA client application (application discovery, automatic installation and update), managing customer accounts (billing and personal information) and implementing basic community features (friends list, voice chat, user invitation).

4.2   Runtime Products

A runtime framework enables application developers to scale their application by distributing it among different hosts and to incorporate and support the advanced Grid functionality provided by the management infrastructure. Two components will be available to them:

- *Real-time framework* [3] is a middleware that provides scalable network communication within Grid systems and sophisticated mechanisms that enable ROIAs to be automatically distributed across multiple servers;

- *Portal* is a scalable request service enabling retrieval of user and session-related information, accessible either with the ROIA client manager or a Web browser.

4.3   Resource Management Products

At the resource management layer, edutain@grid will devise advanced services for automated resource allocation, monitoring, and predicted planning [4] tuned to the requirements of the highly dynamic ROIA. More precisely, the project will have the following outcomes:

- *Resource allocation service* deciding how to map business-oriented requirements to a local resource management policy at a Hoster site. Additionally, this service will aim to facilitate deployment, installation, and update of ROIA servers;
- *Resource monitoring and fault tolerance services* checking the health and availability of resources and detecting potential SLA violations;
- *Capacity management services* predicting future capacity to steer the negotiation strategy for new SLA;
- *Policy management services* maintaining security policies, enabling access rights, and enforcing restrictions for other actors consistent within the terms of existing SLA.

The services in the management layer will act as intermediaries between the business actors and the real-time layer and will enable the protocols for SLA negotiation and steering upon SLA violation.

4.4   Business Products

From the business point of view, edutain@grid will develop services that support economically viable business models, based on a balance of risks and rewards that is attractive to all participants, and supported by security mechanisms and trust models that are cost-effective as well as efficient [5]. The following components will be developed as products:

- *Market service* to support Market Broker operation, allowing offers from Hosters or application/content providers to be matched with requirements from Distributors - this will be a Web service implementation of standard auction models common in the agents community;
- *SLA negotiation services*, allowing the details of an SLA to be agreed with a Hoster – e.g. terms for handling faults or SLA violations, the provision of operational data;
- *User registration services* to support the operation of the ROIA session Coordinator, including business management of the Customer relationship, and access to security token services once business trust in the Customer is established;

– *Accounting services* supporting business-level accounting and micro-payment aggregation, in all the main service providers (Market Broker, Coordinator, Distributor and Hoster), enabling the creation, transfer and aggregation of usage and billing information by other actors.

As a whole, the edutain@grid platform will provide security features allowing cost-effective security, based on business trust relationships negotiated in the business layer, to be propagated across the management and real-time layers. The business layer will therefore support lightweight security procedures based on business trust decisions (e.g. customer credit checks), and capable of supporting pseudonymity and high levels of usability. These will be implemented using conventional WS-Trust services issuing SAML or X.509 security tokens.

5   Conclusions

The edutain@grid framework will offer to end-users unprecedented freedom of action, entertainment, adventure, training, etc. in a virtual world of unique dimensions on top of scalable and dynamic use of compute Grid resources. Facilitating rapid uptake of edutain@grid, the technology will be designed to be generic, scalable, and secure in nature. This will be achieved by providing sophisticated Grid middleware services, distributed real-time computation, and easy to configure user portals. The objective will be to allow both established Grid users and new applications developers to make use of this new technology at minimal cost. At present this has not been achieved in the market place and therefore has the potential to stimulate a whole new community of developers, service providers and end-users.

Within the edutain@grid project two demonstrator applications will be developed and validated including an on-line multi-player game and a multi-user e-learning application in search and rescue (natural environment). The project shall seek to meet the broader market needs through two user groups that will help define the requirements. Membership will be selected from a range of organisations that represent key potential user groups.

References

1. Foster, I., Kesselman, C.: The Grid: Blueprint for a Future Computing Infrastructure, 2nd edn. Morgan Kaufmann, San Francisco (2004)
2. Taylor, I.J., Deelman, E., Gannon, D.B.: Workflows for e-Science. Scientific Workflows for Grids. Springer, Heidelberg (2007)
3. Müller, J., Gorlatch, S.: Rokkatan: scaling an rts game design to the massively multiplayer realm. Computers in Entertainment 4(3), 11 (2006)
4. Siddiqui, M., Villazón, A., Fahringer, T.: Grid allocation and reservation - Grid capacity planning with negotiation-based advance reservation for optimized QoS. In: Supercomputing conference, IEEE Computer Society Press, Los Alamitos (2006)
5. Surridge, M., Taylor, S., Roure, D.D., Zaluska, E.: Experiences with GRIA - industrial applications on a Web services Grid. In: 1st International Conference of e-Science and Grid Computing, pp. 98–105. IEEE Computer Society Press, Los Alamitos (2005)

GridEcon –
The Economic-Enhanced Next-Generation Internet

Jörn Altmann[1,2], Costas Courcoubetis[3], John Darlington[4], and Jeremy Cohen[4]

[1] Intl. University, Bruchsal, Germany
[2] Seoul National University, Seoul, South-Korea
[3] Athens University of Economics and Business, Athens, Greece
[4] Imperial College, London, UK
jorn.altmann@acm.org, courcou@aueb.gr, jd@imperial.ac.uk

Abstract. The major shortcoming of Grid middleware systems is the lack of economic-enhanced Grid services. These new services are necessary in order to let Grid users benefit from the properties of the Grid. Those properties comprise the availability of on-demand computational power, simplicity of access to resources, low cost of ownership, and a pay-for-use pricing model in addition to the already leveraged properties such as cost reduction and aggregated processing power for high-performance computing applications. This paper gives an overview of the EU-funded project GridEcon on Grid economics and business models. It describes its vision of the next generation Grid/Internet, in which individuals, universities, small and medium sized enterprises (SMEs), and large companies have access to the Grid in exactly the same way. Any resource, including servers, storage, software, or data, is accessible as a service. In addition to this, the architecture of an economic-enhanced infrastructure is illustrated and the goal of the project is described.

Keywords: Grid Computing, Grid Economics, Service-Oriented Computing, Economic Modeling, Business Model, Markets, Architecture, and Next-Generation Internet.

1 Introduction

Grid computing has not been commercially taken up to the extent expected during the past few years, although many different (commercial and public domain) Grid middlewares (e.g. glite, Gria, Unicore, Globus, GridBus) have been designed and developed [1][2][3][4][5]. The reason is hidden in the limited leverage of the properties of Grid technology. Currently, enterprises use Grid technology only to consolidate their IT resources, resulting in cost reduction. Only in a few cases, Grid technology is being used for improving the workflow within an enterprise. For example, the combined processing power of geographically distributed servers can be used to reduce the processing time of calculations, or to calculate equations more accurately. It results in reduced time-to-market of products. Grid technology also helps aggregating high-performance computing resources such that applications,

D.J. Veit and J. Altmann (Eds.): GECON 2007, LNCS 4685, pp. 188–193, 2007.

generating more precise results, can be executed on those aggregated resources. The execution of these applications on a single high-performance computer would not work.

However, enterprises miss out on using other properties of Grid technology. These properties comprise the availability of on-demand computational power, simplicity of access to resources, low cost of ownership, and a pay-for-use pricing model. On-demand computational power helps enterprises to deal with unexpected demand economically efficient. Instead of declining a consumer's request simply based on the unavailability of resources (i.e. processing power, storage, bandwidth, software, and data), they could buy those resources on the Grid (if it maximizes the enterprises objective) now. The simplicity of access to resources helps users to access any resource without much effort. Low cost of ownership enables small and medium sized enterprises (SMEs) to get access to resources that they could not afford to purchase as a whole. They only have to pay for the usage of the resources. This model would allow them to compete with large companies, which have the financial resources to buy high-performance computers for their applications.

Considering this situation, two questions arise: First, what is the reason for this low take up of Grid technology; Second, are there no further sustainable business models then those three mentioned above? These questions highlight the need for better understanding the economics behind Grid technology as well as their business models. The GridEcon project addresses these questions [7]. The GridEcon project investigates the economics of participation in a Grid environment as well as how economic principles can be integrated into existing Grid middleware to make it economic-aware. Current Grid middleware lacks these capabilities, as has been analyzed in [8]. A taxonomy of business models has been proposed in [6].

2 Vision of the Future Grid

In a future Grid, which we envision to be the next-generation Internet (i.e Web 3.0), an open market (together with its trading system) is an essential part, where a huge variety of electronic services are traded. Participants (both, consumers and providers) in this market could be anyone from the general public, academia, business, and government, making it a rich economic and social environment. Based on these markets, sustainable Grid business models could be created, offering new ways to generate income. The income could come from customization of information or the creation of new workflows. These new business models would allow participants in the Grid economy to buy services and sell enhanced services at the same time [11].

However, this vision has not been implemented yet. The reason can be found in the fact that there is one technology out of four that is still missing. All of them are necessary to make the vision become reality. The three existing technologies are: service-oriented computing, virtualization of resources, and network computing. Service-oriented computing (e.g. Web services) allows useful capabilities to be encapsulated as easy-to-use, composable services. Hardware virtualization technology allows transparent use of distributed resources. Network computing allows uniform access to the Internet, which is enabled through the convergence of networks and the proliferation of broadband access. The only missing technology is economic-enhanced services, which will give participants in the market tools to evaluate the economic risk and opportunity to engage in a transaction.

This technology will have a significant impact on existing Grid businesses such as location-aware mobile services, consumer advice services, utility computing, brokers, virtual facilities, insurance contracts, software-as-a-service, and information-as-a-service. It will make them accessible to a larger base of customers.

3 Architecture

Looking at the currently available Grid middleware solutions, it becomes obvious that all of the existing Grid middleware solutions do not provide economic-enhanced Grid services. To rectify this situation, the functionality of Grid technology must be enhanced so that an economic-aware operation of Grid services becomes possible. This new functionality would reduce uncertainty and give incentives to end-users not only to consume but also to sell services on the Grid. It could also help stakeholders to resolve their conflicts in preferences. It would, thus, create a new economy, in which all stakeholders can actively participate. An abstract view of this next-generation architecture is shown in the following figure.

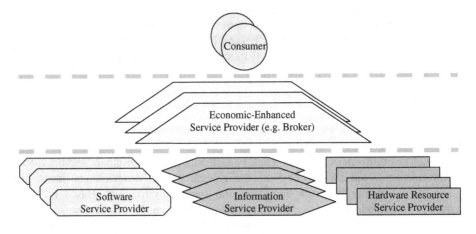

Fig. 1. Architecture of the economic-enhanced next-generation Grid

The Economic-Enhanced Service Provider of Figure 1 will provide tools for trading resources (i.e. software, information, and hardware resources) [10]. It will help Grid stakeholders (i.e. researchers, organizations, companies, and the general public) to deal with the currently existing shortcomings of Grid computing such as risk of relying on outside-company resources, lack of trust, risk in commitment to resource purchases, and uncertainty in capacity planning. These tools, which still need to be developed, range from risk broker services, capacity planning services, to services markets. The risk broker would offer a type of insurance contract to protect against financial loss from unavailable Grid resources or failed Grid resources. An accurate capacity planning tool, which is vital for service provider and end-users, would give support for making decisions about when to purchase new servers, when to put spare resources on the Grid market, and when to buy resources from the Grid. The software services market would allow trading of units of software access. The price of the software access unit would include the price for the software usage and the charge for the hardware resources on which the software would be executed [9]. A hardware resource market will allow selling different server units under a specific pricing scheme. The following figure shows a few examples of hardware resource markets and their relationship to software markets.

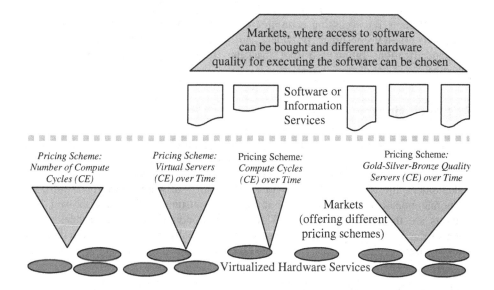

Fig. 2. Architecture of software, information, and hardware resource markets

These kinds of markets, as shown in Figure 2, are the basic services that are needed to make the Grid economic-enhanced. On top of those market services, services as the one mentioned above can be constructed.

There are two major threats. The first threat to Grid computing is the failure of developing and deploying those economic-enhanced Grid services. Since those services also require an open, economic-enhanced architecture for Grid services, which allows any stakeholder to plug in its own services, the second threat to Grid computing is that the Grid community fails to define such an open Grid services architecture.

A market that is based on this architecture will enable collaboration across individual organizational boundaries and reduce the participation risk of Grid stakeholders by allowing economically fair sharing of costs and generated value.

4 Goals of GridEcon

The goals of GridEcon are to twofold. On the one hand, the project has to identify missing technology and software. This comprises the design of the required economic enhancements to Grid technology (as described in the previous section), the implementation of a subset of these service enhancements, as well as the simulation of the workings of the enhancements. On the other hand, the GridEcon project will perform economic and business modeling. It will develop models showing how hardware, software, and information services can be bought and sold on the Grid. It will also investigate potential ecosystems and explore current and future business models.

In particular, the goals of GridEcon are to address the following issues: specify user requiments for accessing economic-enhanced services; SLA composition with respect to pricing; consumer, provider, and service reputation management; service API specification; future and spot markets, insurance contracts, and reservation schemes.

5 Conclusions

This paper discussed the opportunities that come with Grid computing. In particular, it presented the vision of the GridEcon project and the architecture of the future Grid, i.e. the next-generation Internet. The architecture comprises three layers of stakeholders: the basic resource providers (hardware, software, and information); the economic-enhanced service providers; and the consumers. We also showed how markets are the basic building block for other economic-enhanced services in the layer of the economic-enhanced service provider. Finally, we illustrated the different working areas of the GridEcon project and the challenges in this area of research.

Acknowledgement

This work has partially been funded by the European Union in the IST programme "Advance Grid Technology, Systems, and Services" under grant FP6-2005-IST5-033634 with the title GridEcon – Grid Economics and Business Models.

References

1. glite homepage (June 2007), http://glite.web.cern.ch/glite
2. Gria homepage (June 2007), http://www.gria.org
3. Unicore homepage (June 2007), http://www.unicore.eu
4. Globus homepage (June 2007), http://www.globus.org
5. GridBus homepage (June 2007), http://www.gridbus.org
6. Altmann, J., Ion, M., Mohammed, A.: A Taxonomy of Grid Business Models. Gecon2007. In: Intl. Workshop on Grid Economics and Business Models, Rennes, France (August 2007)

7. GridEcon, Grid Economics and Business Models, EU funded project (June 2007), http://www.gridecon.eu

8. Caracas, A., Altmann, J.: A Pricing Service for Grid Computing (2007), at http://it.i-u.de/schools/altmann/?page_id=26

9. Darlington, J., Cohen, J., Lee, W.: An Architecture for a Next-Generation Internet Based on Web Services and Utility Computing. In: ETNGRID2006, Third International Workshop on Emerging Technologies for Next-generation Grid, University of Manchester, Manchester (June 2006)

10. Cohen, J., Lee, W., Darlington, J., McGough, A.S.: A Service-Oriented Utility Grid Architecture Utilising Pay-per-Use Resources. In: COMSWARE2006, Utility Grids 2006 Workshop, 1st International Conference on Communication System Software and Middleware, Delhi, India (January 2006)

11. Cohen, J., Darlington, J., Lee, W.: Payment and Negotiation for the Next Generation Grid and Web. In: Concurrency and Computation: Practice and Experience, pp. 1532–1626. John Wiley & Sons, Ltd, Chichester (2007)

SORMA – Building an Open Grid Market for Grid Resource Allocation

Dirk Neumann, Jochen Stoesser, Arun Anandasivam, and Nikolay Borissov

Institute of Information Systems and Management (IISM)
Universität Karlsruhe (TH)
Englerstr. 14, 76131 Karlsruhe, Germany
lastname@iism.uni-karlsruhe.de

Abstract. The demand for computing and storage resources in a Grid network increases in both academic and industrial application domains. Participants in a network (i.e. companies or research institutes) try to selfishly maximize their individual benefit from participating in the Grid. Setting the right incentives for suppliers and requesters for an efficient usage of the limited Grid resources will motivate the participants to co-operate and provide their idle resources. In this paper we present an economic approach for efficient resource allocation. A market mechanism called *Decentralized Local Greedy Mechanism* [2] satisfies desirable economic properties and thus is deemed promising to enable an efficient allocation of Grid resources.

Keywords: Self-Organizing ICT Resource Management, Open Grid Market, Grid Resource Allocation.

1 Introduction

The vision of a complete virtualization of Information and Communication Technology (ICT) infrastructures by the provision of ICT resources like computing power or storage over Grid infrastructures will make the development of *Open Grid Markets* necessary. Over the Open Grid Market *idle* or *unused* resources (e.g. computational resources) can be supplied (e.g. as services) and client demand can be satisfied not only within one organization but also across multiple administrative domains.

The idea of introducing markets in distributed systems is certainly not new. Despite previous efforts, none of the market-based approaches has made it into practice. The reasons can briefly be summarized by the following three arguments: (1) inadequate market design, (2) insufficient support to use the markets, and (3) improper coupling between the market and state-of-the-art middleware.

The project SORMA (**S**elf-**O**rganizing ICT **R**esource **Ma**nagement, www.sorma-project.org) is funded by the European Union as part of its 6th framework programme. SORMA will design and implement an Open Grid Market in a comprehensive way by addressing all three arguments. Firstly, the economic model provides an economically sound market structure. Secondly, the

D.J. Veit and J. Altmann (Eds.): GECON 2007, LNCS 4685, pp. 194–200, 2007.
© Springer-Verlag Berlin Heidelberg 2007

self-organization model deals with the interaction between the Grid-application and the market by providing intelligent tools. Thirdly, the economic middleware model, which builds the bridge between the self-organization and the economic model on the one side and state-of-the-art Grid infrastructure on the other side. Integrating those three models into the SORMA system, the Open Grid Market is expected to take off in practice and help realizing the benefits of Grid technologies.

The focus of this paper is twofold. In Section 2, we will outline the basic objectives of the SORMA project and the main building blocks of the SORMA system. In Section 3, we will then describe a market mechanism as a specific economic model to illustrate how technical and economic viewpoints integrate in order to determine the allocation of Grid resources to requests as well as the corresponding monetary transfers. Section 4 concludes with a summary and a brief outlook.

2   The Open Grid Market

The overall objective of SORMA is the development of methods and tools for establishing an efficient market-based allocation of ICT resources in order to enable resource accessibility for all users and to increase user satisfaction, profit and productivity. Accounting for both technical and economic issues allows the SORMA methodology to have a strong grip on the technical infrastructure that facilitates market-based resource allocation, while the design of adequate state-of-the-art market mechanisms is based on a solid fundament of economic design knowledge. In addition, SORMA uses concepts from autonomic computing, such that the ICT resource allocation process is being conducted automatically and autonomously. Once the bidding process is virtualized, the ICT system of an organization will self-organize its resource management: Overcapacity will be provided over the market while undercapacity initiates a buying process. A theoretical view on how dynamic market processes work can be found in Neo-Austrian Economics [3]. Since Adam Smith's notion of the *"invisible hand"*, economists describe market participants as competing for limited resources and coordinating themselves through pursuance of their own goals. The goal of market mechanisms is to arrive at a state of coordinated actions – the *"spontaneous order"* – which comes into existence by the community members communicating (bidding) with each other and thus achieving a community (social) goal – being economic efficiency – that no single user has planned for. The main characteristics of the free market self-organization are (1) system elements as utility maximizers, (2) system element strategies that subjectively weigh and choose preferred alternatives in order to reach income or utility maximization, and (3) access to markets as communication platforms to exchange price signals, wrapped in supply and demand offers. Markets and self-organization thus go hand in hand.

The SORMA project combines all those areas of self-organizing ICT resource management in a holistic manner. This is reflected by the high-level architecture, which consists of the following building blocks (Figure 1):

- An Open Grid Market which determines the allocation and corresponding prices in an efficient way and offers in addition complementary market services.
- Economic Grid Middleware which extends common virtualization middleware (such as Globus Toolkit, UNICORE, or gLite) in a way that the allocations determined by the Open Grid Market can be executed.
- Intelligent Tools which automate the bidding process.

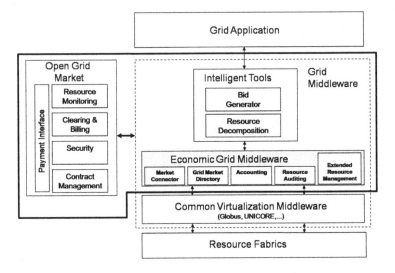

Fig. 1. The SORMA system

In general, a market can be understood as the location where demand and supply meet. The Grid market thus consists of the service consumers and providers, representing demand and supply, and of the software system that implements the market functionalities, representing the Open Grid Market.

Service consumers are those entities that request specific services. For example, a bio-numerical simulation application requests computation and storage services. As it is the goal to have a self-organizing ICT resource management, the service consumer is considered to be the specific Grid application, which provides runtime data of the requester's system. Based on this data, service requests are initiated including detailed service type requirements (e.g. a storage service) and quality of service constraints (e.g. a storage service with at least 300GB free space for four hours). It is assumed that approximate quality and time constraints of the requested services can be determined using prediction models [4].

On the *service provider* side, computer nodes host a set of Grid services (e.g. a computation service or a storage service) and provide information about their current quality characteristics and availability. The node includes a resource manager that provides runtime data about scheduling, allocating, and monitoring of local resources. This resource manager makes use of prediction models in order to determine the approximate future availability of the underlying computational resources.

The *Open Grid Market* component aggregates the bids from service providers and users and subsequently determines an outcome (i.e. allocation of resources and the corresponding prices). Crucial for the working of the Open Grid Market is the clearing mechanism component, which implements the logic how the market is resolved (i.e. auction formats). In the SORMA framework, the clearing mechanism can almost be arbitrarily defined. We will present a sample market mechanism in Section 3.

3   A Decentralized Local Greedy Mechanism

A Grid environment consists of a set of individual users (agents) which we will assume to act rationally and selfishly, meaning the agents intend to maximize their individual benefit from participating in the Grid. In the following, this setting will be introduced as an instance of the class of machine scheduling problems $P|\ r_j\ |\sum w_j C_j$ [1].

Agents submit (computational) tasks to the system, so called jobs. A job j is specified by the tuple $t_j = (r_j, p_j, w_j)$ where $r_j \in \mathbb{R}_0^+$ denotes the job's release date, $p_j \in \mathbb{R}^+$ the job's processing time and $w_j \in \mathbb{R}^+$ the job's weight, i.e. the job's cost of waiting for one additional unit of time. When requesting resources for the execution of a job, the selfish agent[1] j may strategically submit the request (its type [2]) $\tilde{t}_j = (\tilde{r}_j, \tilde{p}_j, \tilde{w}_j) \neq t_j$ to the system. The Grid infrastructure consists of m identical, parallel machines $M = \{1, ..., m\}$. We restrict the strategic analysis of this setting to selfish behavior on the demand side only and consequently assume that machines on the supply side always report true information about their status to the system. Each of the jobs $J = \{1, ..., n\}$ can be executed on any of the m machines. Jobs are assumed to be numbered in order of their release dates, i.e. $j < k \Rightarrow \tilde{r}_j \leq \tilde{r}_k$. Each machine can process one job at a time. Preemption of jobs is not allowed, meaning once a job has been started, it cannot be interrupted externally.

For this setting, Heydenreich et al. [2] propose a Decentralized Local Greedy Mechanism (*DLGM*) which aims at maximizing the agents' overall "happiness" by minimizing the sum of the jobs' weighted completion times $\sum_{j \in J} w_j C_j$ – the costs induced by the system – where C_j denotes j's actual completion time. *DLGM* comprises the following steps:

Step 1: At its chosen release date $\tilde{r}_j$, job j communicates $\tilde{w}_j$ and $\tilde{p}_j$ to every machine $m \in M$. It is assumed that a machine can postpone its release date, therefore $\tilde{r}_j \geq r_j$. The processing time can only be overstated ($\tilde{p}_j \geq p_j$), e.g. by adding

---

[1] In the following we will use the terms "agent" and "job" interchangeably.

unnecessary calculations, but not understated since this could easily be detected and punished by the mechanism [2]. The weight $\tilde{w}_j$ can be reported arbitrarily.

Step 2: Based on the received information $(\tilde{r}_j, \tilde{p}_j, \tilde{w}_j)$, the machines communicate a (tentative) completion time $\hat{C}_j$ and a (tentative) payment $\hat{\pi}_j$ to the job. For obtaining the (tentative) completion time, the remaining processing time of the current job and the processing times of the higher-prioritized jobs in the queue as well as j's own processing time have to be added to the current time. The (tentative) payment equals a compensation of utility loss for all jobs being displaced by j based on j's processing time and the other jobs' waiting cost. The tentativeness is due to the fact that later arriving jobs might overtake job j. This leads to a final (ex-post) completion time $C_j \geq \hat{C}_j$ and a final (ex-post) payment $\pi_j \leq \hat{\pi}_j$ as compensation payments by overtaking jobs might occur (see below). The local scheduling on each machine follows the "weighted shortest processing time first" (WSPT) policy. Jobs are assigned a priority value according to their ratio of weight and processing time: job j has a higher priority than job $k \Leftrightarrow \tilde{w}_j/\tilde{p}_j \geq \tilde{w}_k/\tilde{p}_k$ and is thus inserted in front of k into the waiting queue at the machine.

Step 3: Upon receiving information about its tentative completion time and required payment from the machines, job j makes a binding decision for a machine.

Step 4: Job j is queued at chosen machine i according to its priority and pays $\hat{\pi}_j(i)$ to the lower ranked jobs $L(j)$ at machine i. Compensation in the form of $\tilde{w}_j\tilde{p}_k$ is paid to j by every other job k overtaking j.

DLGM is a promising economic mechanism for scheduling interactive Grid applications due to a number of desirable features. Most importantly, the mechanism has a *polynomial runtime*. It therefore meets the need for immediacy and is especially suitable for an online setting where allocation and payment decisions need to be taken without delays due to computational overhead.

The mechanism's payment scheme is specifically designed for an online setting: payments are computed and transferred to and from jobs as soon as all necessary information is available. Consequently, all monetary exchange is settled when a job leaves the system.

Furthermore, the mechanism achieves *(strongly) budget-balanced payments*. Across the entire set of jobs, all payments paid and received sum up to zero since every incoming job immediately compensates the jobs displaced by it.

A further desirable feature of the mechanism is its *incentive compatibility* on the demand side in dominant strategies. In the restricted strategy space where all jobs $j \in J$ report their true weight w_j, *DLGM* is incentive compatible with respect to the truthful reports of $\tilde{r}_j = r_j$ and $\tilde{p}_j = p_j$. In contrast to $\tilde{r}_j$ and $\tilde{p}_j$, incentive compatibility cannot be guaranteed with respect to a truthful report of $\tilde{w}_j$. In an online setting, from an ex-post perspective with additional information on jobs which have been submitted later on, agents may learn that reporting $\tilde{w}_j \neq w_j$ would have been beneficial [2]. Instead, the concept of a "myopic best response equilibrium" is used in order to show that, at least from an ex-ante perspective, at the time of a job's arrival it is advantageous to give a true report of $\tilde{w}_j$, thus maximizing the tentative utility at time $\tilde{r}_j$.

Given rational behavior of all agents, a *minimal performance can be assured*; the objective value $\sum_{j \in J} w_j C_j$ – which is to be minimized – resulting from the online mechanism where information about jobs is gradually becoming available is compared to the the the optimal solution by an offline mechanism whose computations are based on prior knowledge of the entire information space. Heydenreich et al. [2] derive a value of $\rho = 3.281$ for *DLGM*, meaning the objective value of *DLGM* is guaranteed to be no more than $\rho = 3.281$ times the objective value of an optimal offline mechanism.

4 Conclusion

The efficient allocation of Grid resources requires an adequate scheduling mechanism for matching supply and demand. Markets enable a self-organized adjustment of distributed capacities of resources. The design of the market is crucial for a successful allocation of resources [5]. In this paper the mechanism called *Decentralized Local Greedy Mechanism* is presented which fulfills the economic requirements like incentive compatibility as well as technical requirements like polynomial runtime for determining the allocation. A prominent economic allocation mechanism is market-based proportional share [6]. In this mechanism, the requested service level by a consumer cannot be guaranteed. Other possible mechanisms are the Multi-Attribute Combinatorial Exchange (MACE) by [7] and the Bellagio system [8]. However, they are focused on periodic scheduling whereas in SORMA a continuous mechanism like DLGM is compulsory.

The next step towards supporting a self-organizing management of ICT resources will be the policy-based deployment of business models and user preferences. The requests and bids submitted by the providers and consumer will be supported by intelligent tools to allow a reasonable, automatic trading of resources. Furthermore, the economic middleware conjoins the technical information about the Grid resources with the economic market data. A first implementation based on existing standard interfaces is planned to support WSRF-compliant middleware like Globus. WS-Agreement is used as an information exchange format between the participants and the Open Grid Market. Use cases from industrial consortium members within SORMA will serve so as to verify the SORMA system in practice to realize the vision of a complete virtualization of ICT infrastructures.

References

1. Graham, R.L., Lawler, E.L., Lenstra, J.K., Kan, A.H.G.R.: Optimization and approximation in deterministic sequencing and scheduling theory: a survey. Annals of Discrete Mathematics 5, 287–326 (1979)
2. Heydenreich, B., Müller, R., Uetz, M.: Decentralization and Mechanism Design for Online Machine Scheduling. In: Arge, L., Freivalds, R. (eds.) SWAT 2006. LNCS, vol. 4059, pp. 136–147. Springer, Heidelberg (2006)

3. Hurwicz, L.: The Design of Mechanisms for Resource Allocation. American Economic Review 63(2), 1–30 (1973)
4. Kee, Y.S., Casanova, H., Chien, A.A.: Realistic Modeling and Synthesis of Resources for Computational Grids. In: Proceedings of the ACM/IEEE SC2004 Conference (SC'04)-vol. 00 (2004)
5. Roth, A.E.: The Economist as Engineer: Game Theory, Experimentation, and Computation as Tools for Design Economics. Econometrica 70(4), 1341–1378 (2002)
6. Chun, B.N., Culler, D.E.: Market-based proportional resource sharing for clusters. Technical report, Berkeley, CA (2000)
7. Schnizler, B., Neumann, D., Veit, D., Weinhardt, C.: Trading grid services - a multi-attribute combinatorial approach. European Journal of Operational Research, forthcoming (2006)
8. AuYoung, A., Chun, B., Snoeren, A., Vahdat, A.: Resource allocation in federated distributed computing infrastructures. In: Proceedings of the 1st Workshop on Operating System and Architectural Support for the Ondemand IT InfraStructure (October 2004)

Author Index

Lecture Notes in Computer Science

For information about Vols. 1–4559

please contact your bookseller or Springer

Vol. 4606: A. Pras, M. van Sinderen (Eds.), Dependable and Adaptable Networks and Services. XIV, 149 pages. 2007.

Vol. 4605: D. Papadias, D. Zhang, G. Kollios (Eds.), Advances in Spatial and Temporal Databases. X, 479 pages. 2007.

Vol. 4604: U. Priss, S. Polovina, R. Hill (Eds.), Conceptual Structures: Knowledge Architectures for Smart Applications. XII, 514 pages. 2007. (Sublibrary LNAI).

Vol. 4603: F. Pfenning (Ed.), Automated Deduction – CADE-21. XII, 522 pages. 2007. (Sublibrary LNAI).

Vol. 4602: S. Barker, G.-J. Ahn (Eds.), Data and Applications Security XXI. X, 291 pages. 2007.

Vol. 4600: H. Comon-Lundh, C. Kirchner, H. Kirchner (Eds.), Rewriting, Computation and Proof. XVI, 273 pages. 2007.

Vol. 4599: S. Vassiliadis, M. Berekovic, T.D. Hämäläinen (Eds.), Embedded Computer Systems: Architectures, Modeling, and Simulation. XVIII, 466 pages. 2007.

Vol. 4598: G. Lin (Ed.), Computing and Combinatorics. XII, 570 pages. 2007.

Vol. 4597: P. Perner (Ed.), Advances in Data Mining. XI, 353 pages. 2007. (Sublibrary LNAI).

Vol. 4596: L. Arge, C. Cachin, T. Jurdziński, A. Tarlecki (Eds.), Automata, Languages and Programming. XVII, 953 pages. 2007.

Vol. 4595: D. Bošnački, S. Edelkamp (Eds.), Model Checking Software. X, 285 pages. 2007.

Vol. 4594: R. Bellazzi, A. Abu-Hanna, J. Hunter (Eds.), Artificial Intelligence in Medicine. XVI, 509 pages. 2007. (Sublibrary LNAI).

Vol. 4592: Z. Kedad, N. Lammari, E. Métais, F. Meziane, Y. Rezgui (Eds.), Natural Language Processing and Information Systems. XIV, 442 pages. 2007.

Vol. 4591: J. Davies, J. Gibbons (Eds.), Integrated Formal Methods. IX, 660 pages. 2007.

Vol. 4590: W. Damm, H. Hermanns (Eds.), Computer Aided Verification. XV, 562 pages. 2007.

Vol. 4589: J. Münch, P. Abrahamsson (Eds.), Product-Focused Software Process Improvement. XII, 414 pages. 2007.

Vol. 4588: T. Harju, J. Karhumäki, A. Lepistö (Eds.), Developments in Language Theory. XI, 423 pages. 2007.

Vol. 4587: R. Cooper, J. Kennedy (Eds.), Data Management. XIII, 259 pages. 2007.

Vol. 4586: J. Pieprzyk, H. Ghodosi, E. Dawson (Eds.), Information Security and Privacy. XIV, 476 pages. 2007.

Vol. 4585: M. Kryszkiewicz, J.F. Peters, H. Rybinski, A. Skowron (Eds.), Rough Sets and Intelligent Systems Paradigms. XIX, 836 pages. 2007. (Sublibrary LNAI).

Vol. 4584: N. Karssemeijer, B. Lelieveldt (Eds.), Information Processing in Medical Imaging. XX, 777 pages. 2007.

Vol. 4583: S.R. Della Rocca (Ed.), Typed Lambda Calculi and Applications. X, 397 pages. 2007.

Vol. 4582: J. Lopez, P. Samarati, J.L. Ferrer (Eds.), Public Key Infrastructure. XI, 375 pages. 2007.

Vol. 4581: A. Petrenko, M. Veanes, J. Tretmans, W. Grieskamp (Eds.), Testing of Software and Communicating Systems. XII, 379 pages. 2007.

Vol. 4580: B. Ma, K. Zhang (Eds.), Combinatorial Pattern Matching. XII, 366 pages. 2007.

Vol. 4579: B. M. Hämmerli, R. Sommer (Eds.), Detection of Intrusions and Malware, and Vulnerability Assessment. X, 251 pages. 2007.

Vol. 4578: F. Masulli, S. Mitra, G. Pasi (Eds.), Applications of Fuzzy Sets Theory. XVIII, 693 pages. 2007. (Sublibrary LNAI).

Vol. 4577: N. Sebe, Y. Liu, Y.-t. Zhuang, T.S. Huang (Eds.), Multimedia Content Analysis and Mining. XIII, 513 pages. 2007.

Vol. 4576: D. Leivant, R. de Queiroz (Eds.), Logic, Language, Information and Computation. X, 363 pages. 2007.

Vol. 4575: T. Takagi, T. Okamoto, E. Okamoto, T. Okamoto (Eds.), Pairing-Based Cryptography – Pairing 2007. XI, 408 pages. 2007.

Vol. 4574: J. Derrick, J. Vain (Eds.), Formal Techniques for Networked and Distributed Systems – FORTE 2007. XI, 375 pages. 2007.

Vol. 4573: M. Kauers, M. Kerber, R. Miner, W. Windsteiger (Eds.), Towards Mechanized Mathematical Assistants. XIII, 407 pages. 2007. (Sublibrary LNAI).

Vol. 4572: F. Stajano, C. Meadows, S. Capkun, T. Moore (Eds.), Security and Privacy in Ad-hoc and Sensor Networks. X, 247 pages. 2007.

Vol. 4571: P. Perner (Ed.), Machine Learning and Data Mining in Pattern Recognition. XIV, 913 pages. 2007. (Sublibrary LNAI).

Vol. 4570: H.G. Okuno, M. Ali (Eds.), New Trends in Applied Artificial Intelligence. XXI, 1194 pages. 2007. (Sublibrary LNAI).

Vol. 4569: A. Butz, B. Fisher, A. Krüger, P. Olivier, S. Owada (Eds.), Smart Graphics. IX, 237 pages. 2007.

Vol. 4568: T. Ishida, S. R. Fussell, P. T. J. M. Vossen (Eds.), Intercultural Collaboration. XIII, 395 pages. 2007.

Vol. 4566: M.J. Dainoff (Ed.), Ergonomics and Health Aspects of Work with Computers. XVIII, 390 pages. 2007.

Vol. 4565: D.D. Schmorrow, L.M. Reeves (Eds.), Foundations of Augmented Cognition. XIX, 450 pages. 2007. (Sublibrary LNAI).

Vol. 4564: D. Schuler (Ed.), Online Communities and Social Computing. XVII, 520 pages. 2007.

Vol. 4563: R. Shumaker (Ed.), Virtual Reality. XXII, 762 pages. 2007.

Vol. 4562: D. Harris (Ed.), Engineering Psychology and Cognitive Ergonomics. XXIII, 879 pages. 2007. (Sublibrary LNAI).

Vol. 4561: V.G. Duffy (Ed.), Digital Human Modeling. XXIII, 1068 pages. 2007.

Vol. 4560: N. Aykin (Ed.), Usability and Internationalization, Part II. XVIII, 576 pages. 2007.